Sarah Frank

Key Determinants of Stress Response in Pseudomonas putida KT2440

Sarah Frank

Key Determinants of Stress Response in Pseudomonas putida KT2440

Functional genomics of Pseudomonas putida KT2440 and stress sensitive plasposon mutants

Südwestdeutscher Verlag für Hochschulschriften

Imprint

Any brand names and product names mentioned in this book are subject to trademark, brand or patent protection and are trademarks or registered trademarks of their respective holders. The use of brand names, product names, common names, trade names, product descriptions etc. even without a particular marking in this work is in no way to be construed to mean that such names may be regarded as unrestricted in respect of trademark and brand protection legislation and could thus be used by anyone.

Cover image: www.ingimage.com

Publisher:
Südwestdeutscher Verlag für Hochschulschriften
is a trademark of
Dodo Books Indian Ocean Ltd., member of the OmniScriptum S.R.L Publishing group
str. A.Russo 15, of. 61, Chisinau-2068, Republic of Moldova Europe
Printed at: see last page
ISBN: 978-3-8381-2523-7

Zugl. / Approved by: Hannover, MHH, Diss., 2010

Copyright © Sarah Frank
Copyright © 2011 Dodo Books Indian Ocean Ltd., member of the OmniScriptum S.R.L Publishing group

" Imagination is more important than knowledge. "
" Things should be made as simple as possible, but not any simpler. "

(Albert Einstein)

ACKNOWLEDGEMENTS

Many individuals have provided help during the last years and it is a great pleasure to acknowledge their support.

I would especially like to thank

> my supervisor, Professor Dr. Burkhard Tümmler, for his guidance, advice and support, and especially for his enthusiasm for this project.
>
> my external examiners, Dr. Max Schobert and Prof. Uwe Völker for kindly reviewing this thesis.
>
> the Deutsche Forschungsgesellschaft DFG 653/3 for financially supporting me during the first three years of my PhD.
>
> my collaboration partners for the gentle and patient introduction into their scientific fields; Dr. Frank Schmidt and Dr Manuela Gesell Salazar for processing the proteome samples and supporting me with the data analysis, Christian Jäger for processing the metabolome data and the good times during the "PSYSMO" meetings, and Dr. Christoph Ulmer for enabling the chemostat experiments.
>
> Dr. Robert Geffers and Petra Hagendorf from the Array Facility of the Helmholtz Centre for Infection Research in Braunschweig for their support with the microarrays.
>
> all members of the International Research Training Group: "*Pseudomonas*: Biotechnology and Pathogenicity" for the many fruitful discussions and for the good time.
>
> all members of the Clinical Research Group, both past and present, for their useful advice and all the good moments we shared in and outside the laboratory.
> Special thanks to Jens for his great support in the set-up of experiments and all the hours of discussions, Lutz, who always had an open ear for any kind of problems, Colin for his advice in bioinformatical questions and his support in the transcriptome analysis, and to Anya helping me with the metabolome analysis.
>
> Jens and Colin for reading all these pages.
>
> Sonja and Nina for not just being the kindest lab-mates but also for their encouragement and motivation and most important their friend-ship.
>
> my family: my parents Ingrid and Winfried, my sister Berrit and my boyfriend Ocke for their constant support, encouragement, love and understanding which had led me this far. Thank you ever so much.

ABSTRACT

Pseudomonas putida KT2440 is a biosafety strain, which has retained its capability to survive and function in the environment. Its broad metabolic capability to degrade a broad spectrum of aromatic compounds and still grow at temperatures as low as 5°C makes it a good candidate as a model organism to study genome wide responses to changing growth conditions which mimic environmental changes in natural habitats.

In this study a functional genomics approach was used to examine the stress response of *P. putida* KT2440 wild type and five cold sensitive Tn*5* mutants (with plasposons inserted in *cbrA*, *cbrB*, *pcnB*, *vacB* and *bipA*) to cold shock by decreasing the temperature from 30°C to 10°C. For this, transcriptome data derived from three different transcriptome platforms (Illumina cDNA sequencing, Affymetrix microarrays and Progenika microarrays), as well as metabolome and proteome data were combined. The wild type and a *pcaI*::Tn*5* mutant were furthermore exposed to high concentrations of sodium benzoate (45 mM).

The benzoate stress experiments revealed a two-phase response, where initially genes involved in degrading and extruding benzoate are induced to reduce the benzoate concentration within the cell and subsequently genes associated with cell wall stress are induced, which reflects repair mechanisms of the cell membrane. The analysis of a *pcaI*::Tn*5* mutant, which is knocked-out for the degradation of benzoate via the ß-ketoadipate pathway, showed that partial deletion of the pMB1 *oriR* from the plasposon of *pcaI*::Tn*5* isolates, could rescue growth on benzoate and exhibited that transposons can undergo secondary mutations under high selective pressure.

The comparison of the three transcriptome platforms showed that Illumina cDNA sequencing is a promising alternative for transcriptome analysis, since it overcomes limitations of microarrays, such as signal saturation and dependence on selection of probes from predefined sequences. The transcriptome data derived from Illumina cDNA sequencing revealed many novel protein coding ORFs and transcripts encoding potential sRNAs in intergenic regions which were missed in the initial annotation.

The transcriptome analyses revealed 159 consistently differentially regulated genes. Following cold shock intermediary metabolism was down-regulated, as indicated by many genes involved in tricarboxylic acid cycle and amino acid metabolism, which were less expressed at 10°C. In contrast, a predominant number of hypothetical proteins and many transcripts identified in the intergenic regions according to the Illumina cDNA sequencing results were induced upon cold shock. This demonstrated that little is known about the active cold shock response in *P. putida*. The complementary proteomics approach revealed additional proteins involved in membrane maintenance and translation efficiency. Furthermore, it was shown that posttranscriptional modifications play a major role in adaptation processes to low temperatures.

According to the metabolome data, phosphorylated sugars and amino acids, which are close to the citric acid cycle, constitute the core metabolome of *P. putida* and are essential for cold shock response. The five Tn*5* mutants were strongly affected in the metabolism of complex amino acids and sugars of the pentose phosphate pathway, which leads to cold sensitivity. Though the mutants are affected in genes of diverse functionality, the transcriptome and metabolome data showed

ABSTRACT

similar cold shock profiles, and are mainly affected in metabolism of complex amino acids. For instance, the *ped* cluster, which is involved in phenylalanine metabolism, was found to be consistently repressed in all five mutants. Furthermore, the consistent phenotype in the growth experiments and the similar transcriptome and metabolome profile of *cbrA*, *cbrB*, *pcnB*, *vacB* and *bipA* with *cbrB* being most distinct, constitutes further evidence for the global regulatory role of *cbrB* and supports the hypothesis that *cbrB* is the link between carbon metabolism, mRNA degradation and translational efficiency.

KURZFASSUNG

Bakterien der Spezies *Pseudomonas putida* überleben in verschieden Habitaten, unter anderem durch die Nutzung vieler verschiedener Kohlenstoffquellen und schnelle Anpassung an sich verändernde Umweltbedingungen. *P. putida* ist z. B. in der Lage, viele aromatische Kohlenwasserstoffe umzusetzen oder vermehrt sich auch bei niedrigen Temperaturen, und ist somit ein geeigneter Modellorganismus für genomweite Analysen der bakteriellen Antwortmechanismen auf veränderte Umweltbedingungen.

In dieser Arbeit wurde mit „omics"-Analysen für den Referenzstamm KT2440 und kältesensitive Transposonmutanten (*cbrA*-, *cbrB*-, *pcnB*-, *vacB*- und *bipA*-Mutanten) die Reaktion auf eine Temperatursenkung von 30°C auf 10°C untersucht (Kälteschock). Dazu wurden Transkriptom-Daten mit drei verschiedenen Systemen generiert und verglichen (Illumina cDNA-Sequenzierung, Affymetrix Mikorarrays, Progenika Mikroarrays) sowie Metabolom- und Proteom-Profile erstellt. Für KT2440 und eine *pcaI*-Mutante wurde auch die Stressantwort bei Erhöhung der Benzoatkonzentration im Wachstumsmedium analysiert.

Diese Untersuchungen zeigen eine in zwei Phasen ablaufende Antwort auf Benzoat-Stress: Zuerst wurden Gene induziert, die über Abbau oder Ausschleusung die intrazelluläre Benzoatkonzentration reduzieren. In der zweiten Phase wurden die Struktur und Zusammensetzung der Zellwand dem exogenen Benzoatstress angepasst. Bei der Mutante *pcaI*::Tn5 wurde ein sekundäres Mutationsereignis beobachtet. Die bei Benzoat-Präsenz letale Plasposon-Insertion in *pcaI* wurde durch partielle Deletionen dieser Sequenz wieder kompensiert. Dies verdeutlichte, dass auch in Transposon-Konstrukten bei hohem selektiven Druck bisher nicht beschriebene Sekundärmutationen auftreten können.

Bei der Auswertung der Kältestress-Transkriptom-Daten wurden die technischen Vorteile der Illumina-cDNA-Sequenzierung gegenüber den Mikroarray-Systemen deutlich. Da bei der Sequenzierung auch Transkripte von schwach exprimierten Genen detektiert werden konnten und zudem die Anzahl detektierbarer Transkripte nicht durch vorherige Selektion limitiert war, konnte insgesamt eine deutlich höhere Anzahl an exprimierten Genen detekiert werden. Außerdem wurden viele Transkripte identifiziert, die bisher als „intergenisch" definierten Regionen zugeordnet werden konnten, aber tatsächlich Proteine oder sRNAs kodieren.

In allen Transkriptomanalysen ließen sich 159 bei 30°C und 10°C unterschiedlich exprimierte Gene nachweisen. Demnach wurden in *P. putida* als Antwort auf den Kältestress viele Gene des Intermediärstoffwechsels herunterreguliert, insbesondere am Aminosäurestoffwechsel und am Citratzyklus beteiligte Gene. Im Proteom erhöhte sich bei 10°C die Abundanz von Regulatoren der Membranfluidität und Elemente des Translationsapparates. Diese Ergebnisse verdeutlichten die zentrale Bedeutung von posttranskriptionellen Modifikationen als Teil der Stressantwort.

Die Metabolom- und Transkriptom-Profile der kältesensitiven Mutanten zeigten, verglichen mit dem Wildtyp, eine deutliche Einschränkung der Umsetzung aromatischer Aminosäuren (erkennbar an einer Repression des *ped*-Genclusters), die demnach essentieller Bestandteil der Kältestressantwort ist. Die konsistenten Phänotypen der *cbrA*-, *pcnB*-, *vacB*- und *bipA*-Mutanten, und zum Teil auch der *cbrB*-Mutante, wiesen darüber hinaus darauf hin, dass die entsprechenden

Gene eine regulatorische Einheit mit *cbrB* als globalem Regulatorgen bilden, über die der zentrale Stoffwechsel, mRNA-Stabilisierung und Translationseffizienz miteinander verknüpft werden.

TABLE OF CONTENTS

ABBREVIATIONS 13
1 INTRODUCTION 15
1.1 *Pseudomonas putida* 15
1.2 Key Factors in Bacterial Stress Response 19
1.3 Gene Expression Analysis 25
1.4 Systems Biology 29
1.5 Objectives of the Present Study 31
2 MATERIALS AND METHODS 33
2.1 Materials 33
   2.1.1 Equipment 33
   2.1.2 Consumables 33
      2.1.2.1 *Pseudomonas putida* Genome Oligonucleotide Array 33
      2.1.2.2 Affymetrix microarray 34
      2.1.2.3 Consumables 34
   2.1.2 Chemicals 35
   2.1.3 Enzymes 36
2.2 Media and Solutions 37
   2.2.1 Media 37
   2.2.2 Solutions 38
2.3 Biological Materials 45
   2.3.1 Strains 45
   2.3.2 Plasmids 45
   2.3.3 Oligonucleotides 46
2.4 BioFlo 110 Modular Benchtop Fermenter 47
2.5 Microbiological Methods 49
   2.5.1 Bacterial growth conditions 49
   2.5.2 Determination of cell density 49
   2.5.3 Determination of colony forming units 49
   2.5.4 Determination of cell dry weight 49
   2.5.5 Maintenance of bacterial cultures 50
   2.5.6 Growth and purification of plasposon mutants 50
   2.5.7 Genetic complementation 50
      2.5.7.1 Generation of electrocompetent cells 51
      2.5.7.2 Electrotransformation 51
   2.5.8 Motility assays 51
      2.5.8.1 Swimming 51
      2.5.8.2 Swarming 51
      2.5.8.3 Twitching 52

2.6 Molecular Biology Methods 53
2.6.1 Isolation of DNA 53
2.6.1.1 Isolation of genomic DNA from *Pseudomonas putida* 53
2.6.1.2 Isolation of plasmid DNA 53
2.6.2 Quantification of nucleic acids 54
2.6.3 Polymerase Chain Reaction 54
2.6.3.1 Standard PCR 56
2.6.3.2 Combinatory colony hot-start PCR 57
2.6.4 Agarose gel electrophoresis 58
2.6.5 Isolation of DNA fragments from agarose gels 59
2.6.6 DNA preparation for genetic complementation 59
2.6.6.1 Restriction digestion of DNA 59
2.6.6.2 Ligation 60
2.6.7 DNA:DNA Hybridization 60
2.6.7.1 Generation of digoxigenin-labelled DNA probes 60
2.6.7.2 Restriction digestion of genomic DNA 61
2.6.7.3 Transfer and fixation of DNA to a membrane (Southern Blot) 61
2.6.7.4 Hybridization and detection of digoxygenin-labelled DNA 62
2.6.7.5 Regeneration of hybridized DNA membranes 62
2.6.8 The Y-linker method 63
2.6.8.1 Generation of the Y-linker 63
2.6.8.2 Y-linker Ligation 64
2.6.9 DNA sequencing 65
2.7 RNA Working Technique and Transcriptome Analysis 67
2.7.1 RNA handling 67
2.7.2 RNA extraction and storage 67
2.7.3 Formaldehyde agarose gel electrophoresis 68
2.7.4 cDNA generation 68
2.7.5 RT/PCR 69
2.7.6 cDNA labelling 70
2.7.6.1 cDNA labelling for *Pseudomonas putida* Genome Arrays (Progenika) 70
2.7.6.2 cDNA fragmentation and labelling for Affymetrix microarrays 71
2.7.7 Microarray blocking and hybridization 72
2.7.7.1 Blocking and hybridization for *Pseudomonas putida* genome arrays (Progenika) .. 72
2.7.7.2 Hybridization of Affymetrix microarrays 73
2.7.8 Microarray data elevation 73
2.8 Illumina cDNA Sequencing 75
2.9 Metabolome Analysis 77
2.10 Proteomic Analysis by LC-ESI-MS/MS 79

3 RESULTS AND DISCUSSION ... 81

3.1 Verification of Tn5 Plasposon Mutants ... 83

3.1.1 Determination of plasposon insertion sites via combinatorial colony hot-start PCR ... 85
3.1.2 Identification of the correct plasposon insertion site in mutant 12E11 ... 87
3.1.3 Phenotypic verification of stress sensitive mutants ... 89
3.1.4 Complementation in *trans* ... 91

3.2 Growth Experiments under Stress Conditions ... 95

3.2.1 Growth Characteristics under Increasing Sodium Benzoate Concentrations ... 96
3.2.2 Comparison of growth under cold adaptation and cold shock ... 98

3.3 Pseudomonas putida pcaIJ Plasposon Rescue ... 103

3.3.1 Wild type KT2440 outcompeted a 10^6 fold excess of an isogenic *pcaIJ* plasposon mutant ... 104
3.3.2 Partial deletion of plasposon insertion rescues *pcaIJ* activity ... 104

3.4 Stress Induction by Benzoate Pulse Implementation ... 109

3.5 Deep RNA Sequencing ... 115

3.5.1 Evaluation of alignment quality and transcript coverage ... 116
3.5.2 Low and high abundant transcripts ... 117
3.5.3 Finding unpredicted ORFs, small RNAs and unclassified transcripts in intergenic regions ... 122

3.6 Comparison of Three Transcriptome Platforms ... 141

3.6.1 Correlation between the results from three transcriptome platforms ... 143
3.6.2 Gene expression profiles derived from the three transcriptome platforms ... 153

3.7 Key Players in Cold Shock Stress of Pseudomonas putida KT2440 ... 163

3.7.1 Transcriptome analysis ... 163
3.7.2 Proteome analysis ... 180

3.8 The Transcriptome Profile of Cold Sensitive Tn5 Mutants ... 191

3.9 Metabolome Profile of Pseudomonas putida KT2440 Wild Type and Stress Sensitive Tn5 Mutants ... 213

4 PERSPECTIVES ... 233

5 CONCLUSION ... 235

6 REFERENCES ... 241

ABBREVIATIONS

A	absorption	m	milli (10^{-3}); meter
aa	amino-allyl	M	molar; Mega (10^9)
amol	attomol	Mb	Megabase
AT	adenosine and thymine	MES	morpholinoethanesulfonic acid
ATP	adenosine triphosphate	min	minute(s)
bp	base pair	MM	mismatch
BSA	Bovine Serum Albumin	MOPS	morpholinopropanesulfonic acid
c	concentration	mRNA	messenger RNA
CDP	2-chlor-5-(4-methoxyspiro{1,2-dioxetan-3,2'-(5'-chlor) tricyclo[3.3.1.1.3,7]Decan}-4-yl)-1-phenylphosphate	MS	mass spectrometry
cfu	colony forming units	MSTFA	N-methyl-N-trifluoroacetamide
CIP	calf intestine phosphatase	µ	micro (10^{-6}); growth rate (h^{-1})
cDNA	complementary DNA	μ_{max}	maximal growth rate (h^{-1})
cm	centimeter	n	nano (10^{-9})
cy3 (5)	cyanine3 (5)	NEB	New England Biolabs
ddH$_2$O	double distilled H$_2$O	NJ	New Jersey
DEPC	diethylpyrocarbonate	OD	optical density
DIG	digoxigenin	ORF	open reading frame
DMSO	dimethylsulfoxid	p	pico (10^{-12}); probability
DNA	deoxyribonucleic acid	PCR	polymerase chain reaction
DNase	deoxyribonuclease	PCU	Packet Control Unit
dNTP	deoxynucleotide triphosphate	PHA	Polyhydroxyalkanoate
dO$_2$	dissolved oxygen	PM	perfect match
DPFC	Digital Pressure and Flow Control	ppm	parts per million
dUTP	deoxyuridine triphosphate	*P. putida*	*Pseudomonas putida*
E. coli	*Escherichia coli*	PVP	Polyvinilpyrolidone
EDTA	ethylenediaminetetraacetic acid	RNA	ribonucleic acid
e.g.	for example	RNase	ribonuclease
et al.	et alteri (and others)	RPKM	reads per kilobase and million
EtOH	ethanol	rpm	revolutions per minute
exp.	exponential	rRNA	ribosomal RNA
Σ	extinction	RT	room temperature; reverse transcription
F		SAPE	Streptavidin-phycoerythirin
Fig.	figure	SDS	sodium dodecyl sulfate
FDR	false discovery rate	sec	second(s)
g	gramm; g-force	sRNA	small RNA
G	Giga (10^{12})	SSC	standard saline citrate
Gb	Gigabase	SSPE	Saline-Sodium Phosphate-EDTA
GC	guanine and cytosine	t_{den}	denaturation time
GC-MS	gas chromatography-mass spectrometry	t_{elong}	elongation time
h	hour; height	T_{ann}	annealing temperature
HEPES	4-(2-hydroxyethyl)-1-piperazineethanesulfonic acid	T_M	melting temperature
hrs	hours	TBE	Tris/Borate/EDTA
HZI	Helmholtz Zentrum für Infektionsforschung	TE	Tris-EDTA
k	kilo	Tris	Tris(hydroxymethyl)aminomethane
kb	kilobase	U	unit (unit of enzymatic activity)
l	dilution / layer thickness	UPCL	Ultra Performance Liquid Chromatography
L	liter spectrometry	USA	United States of America
LB	Luria-Bertani	UV	ultraviolet
LC-ESI-MS/MS	Liquid chromatography electrospray ionisation tandem mass	V	voltage
ln	natural logarithm	W	watt
log	logarithmic	% v/v	percentage volume per total volume
LOWESS	locally weighted scatterplot smoothing	% w/w	percentage by weight per total volume
LTQ-FTICR	Hybrid Linear Ion Trap Fourier Transform Ion Cyclotron Resonance		

1 INTRODUCTION

1.1 *Pseudomonas putida*

The genus *Pseudomonas* belongs to the γ-subclass of the Proteobacteria. To date, more than 120 *Pseudomonas* species have been described, including fluorescent Pseudomonads, like *Pseudomonas putida*, *Pseudomonas fluorescens* and *Pseudomonas aeruginosa*, phytopathogenic species such as *Pseudomonas syringae* and non-fluorescent Pseudomonads such as *Pseudomonas stutzeri* and *Pseudomonas mendocina* (Peix *et al.*, 2009). Information and annotations on 17 sequenced *Pseudomonas* genomes is stored at the *Pseudomonas* Genome Database (http://www.pseudomonas.com).

The species *P. putida* is a non-pathogenic, saprophytic bacterium that is frequently isolated from soil and water environments, and is known for its ability to degrade a wide variety of organic compounds, including xenobiotics such as aliphatic and aromatic hydrocarbons. Many studies demonstrated its metabolic versatility and capability for bioremediation, since *P. putida* is not only able to degrade these compounds, but is in most cases capable of using these as sole carbon and energy sources. These include toluene (Inoue *et al.*, 1991), benzene and ethylbenzene (Parales *et al.*, 2000; Baldwin *et al.*, 2000), styrene (Okamoto *et al.*, 2003; Dunn *et al.*, 2005), xylene and naphthalene (Phoenix *et al.*, 2003) benzoate and phenol (Feist & Hegemann, 1969a, 1969b; Janke *et al.*, 1981; Ramos *et al.*, 1995). Furthermore, *P. putida* exhibits resistance to even very high concentrations of aromatic hydrocarbons (Segura *et al.*, 2005; Reva *et al.*, 2006).

In environmental *P. putida* strains, enzymes for the metabolism of aromatic hydrocarbons are often encoded on plasmids that account for a major part of their biodegradative potential (Greated *et al.*, 2002; Nelson *et al.*, 2002) and channel the catabolized substrates into the central metabolism. Most of the aromatic compounds are metabolized via catechol, protocatechuate and 4-hydroxybenzoate that enter the ß-ketoadipate pathway, where the aromatic ring structure is cleaved, and finally funnel into the citric acid cycle (Harwood & Parales, 1996). Beside its great potential for degradation of pollutants, *P. putida* has to cope with various abiotic stresses in its environment, such as nutrient limitations, temperature shifts, oxygen and water stress (Hecker & Völker, 2001; Hallsworth *et al.*, 2003; Angelis & Gobetti, 2004; Phadtare, 2004).

INTRODUCTION

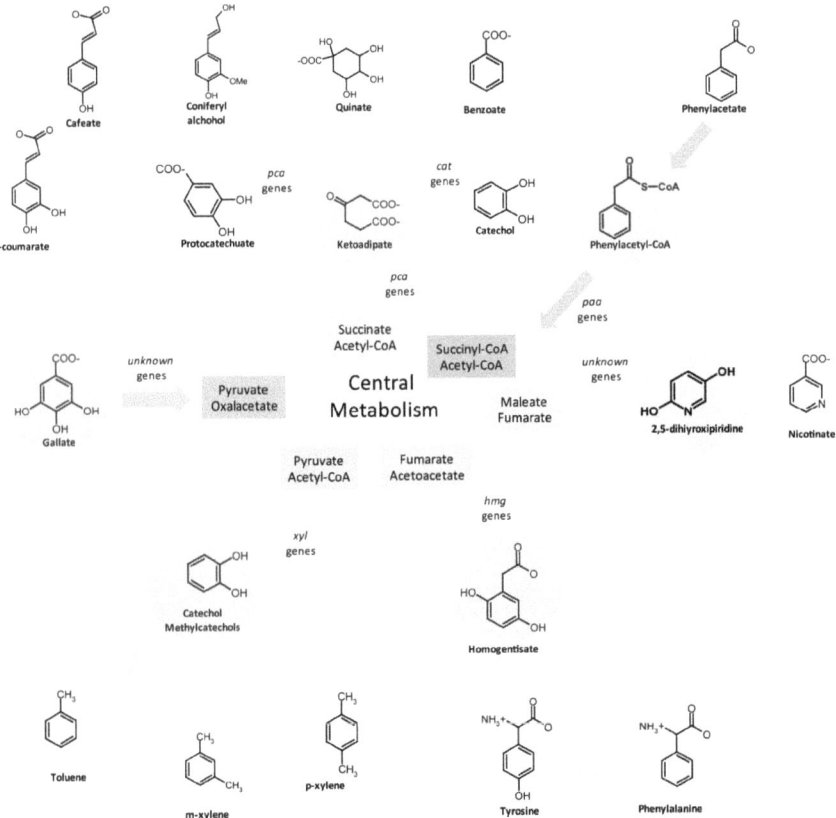

Fig. 1.1 General depiction of the aromatic compound degradation routes present in *P. putida*. The protocatechuate (*pca* genes) and catechol (*cat* genes) branches of the beta-ketoadipate pathway are shown as well as peripheral pathways by orange arrows. The homogentisate pathway (*hmg* genes) is represented by green arrows and the phenylacetate pathway (*paa* genes) is represented by purple arrows. The nicotinate and gallate pathways (unknown genes) are shown by green and red arrows, respectively. Finally, the Tol pathway (*xyl* genes from pWW0 plasmid) for toluene and xylene degradation is represented by blue arrows. The initial aromatic compounds are indicated by green circles and the central metabolic compounds for each pathway are also highlighted. (Nogales *et al.*, 2008).

Pseudomonas putida KT2440

The lab strain *P. putida* KT2440 (Bagdasarian *et al.*, 1981) emerged from the *Pseudomonas arvilla* mt-2 strain, originally isolated in Japan and reclassified to *Pseudomonas putida* mt-2 (Nakazawa & Yokota, 1973; Williams & Murray, 1974), which harbours the TOL plasmid pWW0 encoding the specific pathway for toluene and xylene degradation via the meta-cleavage branch of the ß-ketoadipate pathway (Williams & Murray, 1974; Worsey & Williams, 1975). Though, the strain KT2440 lacks the TOL plasmid, it is capable of degrading aromatic compounds via the ortho-cleavage branch of the ß-ketoadipate pathway.

KT2440 has been declared a biosafety strain, GRAS (**G**enerally **R**ecognized **A**s **S**afe), that has retained its ability to survive and function in the environment. These characteristics make *P. putida* KT2440 a promising candidate as model organism that can be used as host-strain for containment systems for release in the environment and applications in biotechnology (Nüsslein *et al.*, 1992; Jensen *et al.*, 1993; Molina *et al.*, 1998).

Indeed, *P. putida* strains have been shown to be suitable for the production of chemicals from natural renewable resources. Wierckx *et al.* (2005) engineered a solvent tolerant *P. putida* S12 strain for the bioproduction of phenol from glucose. Similarly, a *P. putida* S12 strain was constructed for the production of fine chemicals cinnamic acid from glucose or glycerol (Nijkamp *et al.*, 2005) and 3-methylcatechol (Wery *et al.*, 2000). Furthermore, it was shown that *P. putida* is suitable for heterologous expression of secondary metabolites derived from myxobacteria that produce products of medical and industrial importance such as antibiotics and drugs against cancer (Gross *et al.*, 2005; Wenzel *et al.*, 2005; Gross *et al.*, 2006a, 2006b).

These examples emphasize that *P. putida* KT2440 as a biosafety strain is an ideal candidate for research in the design for production strains that can be used for industrial applications.

1.2 Key Factors in Bacterial Stress Response

In the following five genes will be described that have been identified in several bacterial genera to be important for cell physiology and stress response to changing environmental conditions. In particular, Reva *et al.* (2006) have shown that these genes are involved in benzoate and cold stress response in *P. putida* KT2440 which is the objective of this study.

The CbrA-CbrB two-component system

Two-component systems are regulatory systems that allow the cells to sense and response to environmental changes. These signal transduction systems consist of a histidine protein kinase and a response regulatory protein. In general, histidine kinases sense environmental stimuli resulting in their activation and thus initiating transfer of a phosphoryl group to the response regulator. Due to the phosphotransfer, downstream mechanisms are then again activated displaying the specific response. Hence, these systems play an important role in the capacity of bacteria to adapt to a wide range of environmental changes, such as nutrient availability, osmolarity and oxygen stress (Hoch & Silhavy, 1995).

In the phytopathogen *Erwinia chrysanthemi* 3937 CbrAB was shown to be involved in virulence by negatively regulating chrysobactin biosynthesis in the presence of iron (Expert *et al.*, 1992). Studies on *Sinorhizobium meliloti*, which establishes a nitrogen-fixing symbiosis with leguminous plants, revealed a CbrA-dependent regulation of an inner ABC transporter required for lipopolysaccharide transport. Succinoglycan, an exopolysaccharide, is crucial for facilitation of symbiosis between this bacterium and its host plant (Gibson *et al.*, 2006). Furthermore, a gene expression analysis of a *cbrA*::Tn5 mutant revealed many differentially regulated genes involved in cell wall biogenesis, motility and chemotaxis (Gibson *et al.*, 2007). In accordance with their previous findings, the authors hypothesized that CbrA plays a role in a developmental switch during symbiosis since nine genes that are known to be involved in the bacterial invasion of its host were identified to be regulated by CbrA.

In Pseudomonads the CbrA-CbrB two-component regulatory system was described to be involved in many regulatory processes that control the expression of several catabolic pathways. In coordination with NtrC, the nitrogen control system activator, it ensures the intracellular nitrogen:carbon balance (Nishijyo *et al.*, 2001) and controls the utilization of many carbon and nitrogen sources such as mannitol, glucose, pyruvate and citrate, several amino acids (arginine, histidine and proline) and polyamines (Nishijyo *et al.*, 2001; Li & Lu, 2007; Zhang & Rainey, 2008). For *Pseudomonas aeruginosa* it was furthermore shown that an impairment of cytotoxicity caused by an imbalanced type III secretion system could be complemented by mutation insertion in *cbrA*. This also resulted in overexpression of *cbrB* (Rietsch *et al.*, 2004).

All these findings indicate that the two-component system CbrAB is a highly ranked regulatory system in Pseudomonads as well in other genera. Sonnleitner *et al.* (2009) supported the central role of CbrAB by detecting the small RNA CrcZ and revealing it as a global regulator of carbon catabolite repression. *CrcZ* is located downstream of *cbrB* and its expression is dependent on CbrAB. The analysis of *crcZ* expression during growth on different carbon sources demonstrated the regulation of catabolite repression by CrcZ and hence of other genes under Crc control such as *amiE* and *benR*. They predicted a model for the CbrAB-CrcZ-Crc system similar to the organization of the Gac/Rsm pathway (Lapouge *et al.*, 2008) where the two-component system regulates the expression of sRNAs that sequester RNA-binding proteins and therefore induce expression of target genes.

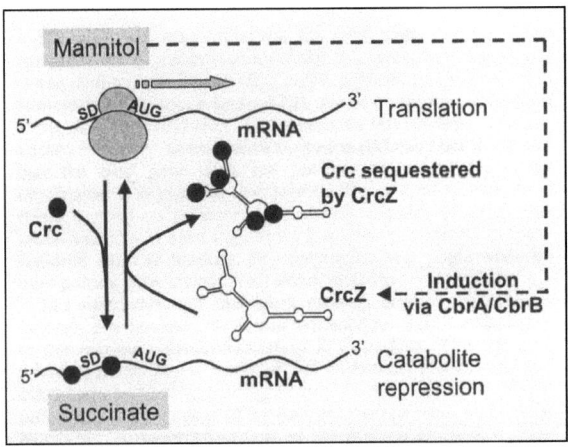

Fig. 1.2 Model of CrcZ as an antagonist of Crc in catabolite repression. The concentration of CrcZ changes according to the carbon source. In the presence of a preferred carbon source (e.g., succinate), the level of CrcZ is low and Crc binds to catabolite repression-sensitive mRNAs such as *amiE* mRNA and thereby blocks ribosome binding. When a non preferred substrate source such as mannitol is the sole carbon source, the expression of CrcZ sRNA increases under the control of the CbrA/CbrB two-component system. This results in sequestration of Crc protein by CrcZ and allows ribosome binding and translation of the target mRNAs. With glucose as the sole carbon source, an intermediate amount of CrcZ allows partial sequestration of Crc protein, leading to moderate expression of target mRNAs (adapted from Sonnleitner *et al.*, 2009).

The poly(A) polymerase PcnB

Bacterial mRNAs can be polyadenylated at their 3' end resulting in a poly(A) tail which is about 30 nucleotides long (Anantharaman *et al.*, 2002). The poly(A) tails promote the degradation of mRNAs since they allow the binding of RNA degrading enzymes to the polyadenylated tail. The secondary structure would otherwise block the binding to the mRNA's 3' end (O'Hara *et al.*, 1995; Régnier & Arraiano, 2000). Though polyadenylation mainly promotes mRNA degradation, it is also thought to play a role in mRNA stabilization and translation under certain conditions (Sarkar, 1997).

The poly(A) polymerase *pcnB* was first identified in *Escherichia coli* where it was found to control plasmid copy numbers and furthermore to be involved in cell growth as a mutation in *pcnB* resulted in a 67% reduction of growth rate in rich medium (Lopilato *et al.*, 1986; Liu & Parkinson, 1989). Cao & Sakar (1992) showed that *pcnB* is a principle poly(A) polymerase and O'Hara *et al.* (1995) disclosed that it is responsible for the polyadenylation of over 90% of cellular mRNA.

A recent study on *pcnB* in *Pseudomonas fluorescens* revealed its CbrAB-dependent regulation (Zhang *et al.*, 2010) with transcription starting from a σ^{70} type promoter rather than from the σ^{54} promoter as previously thought. Though the σ^{54} promoter is also regulated by CbrB, it controls the expression of the small RNA CrcZ (Sonnleitner *et al.*, 2009). In addition, growth experiments by testing various substrates as sole carbon source showed that *pcnB* deletion resulted in growth rate reduction when grown on certain substrates such as glycerol and succinate and a β-galactosidase assay indicated a strong correlation between growth rate and the level of *pcnB* transcript. Thus, they demonstrated the central regulatory role of CbrB and concluded that polyadenylation is important for bacterial acclimation to changing environmental conditions and that CbrAB and PcnB display a link between mRNA degradation and carbon metabolism. The functional interaction of CbrAB and PcnB was furthermore supported by their simultaneous identification of being involved in stress response to cold temperatures in *P. putida* (Reva *et al.*, 2006).

The exoribonuclease R VacB

RNase R is generally involved in posttranscriptional regulation of mRNA stability (Andrade *et al.*, 2006), but was also found to be involved in RNA quality control (Li *et al.*, 2002) and decay (Cheng & Deutscher, 2005). Furthermore, *vacB* was associated with the expression of virulence genes in various organisms: *Shigella flexneri* and *E. coli* (Tobe *et al.*, 1992; Cheng *et al.*, 1998), *Helicobacter pylori* (Tsao *et al.*, 2009) and *Aeromonas hydrophila* (Erova *et al.*, 2008). In *Helicobacter pylori*, the gene expression profile of a *vacB* showed furthermore an altered expression in response to changing growth conditions such as pH and temperature. Thus, it is likely that VacB in *H. pylori* has various regulatory functions. The *vacB* gene in *Aeromonas hydrophila* was furthermore required for growth at low temperatures. The role of RNase R as cold shock protein was discussed in another study (Cairrão *et al.*, 2003) where its expression was regulated by temperature and RNase R was seven to eight-fold induced upon cold shock. In *P. putida*, the *vacB* gene contains a S1 RNA binding domain that exhibits structural similarity to cold shock proteins. Indeed, a *vacB*::Tn*5* mutant was impaired in growth at low temperatures (Reva *et al.*, 2006). A more global role of RNase R was indicated by a study on catabolite repression control of BkdR in *P. putida* and *P. aeruginosa* (Hester *et al.*, 2000). BkdR is the transcriptional activator of the *bkd* operon encoding a multienzyme complex branched-chain keto acid dehydrogenase (Madhusudhan *et al.*, 1993, 1997) and was shown to be regulated by Crc. A transposon mutant analysis revealed three mutants affected in BkdR regulation, two of which were *crc* mutants and one was a *vacB* mutant, indicating that VacB might also be involved in the regulation of some carbon utilization pathways. MacGregor *et al.* (1996) already showed that *crc* has some sequence similarity to exonucleases.

The GTPase BipA

BipA is a highly conserved GTPase that exhibits several ribosome-associated cellular functions that are related to stress response (Farris *et al.*, 1998; Pfennig & Flower, 2001; DeLivron & Robinson, 2008; DeLivron *et al.*, 2009). In *E. coli bipA* was described to be involved in virulence regulation since it controls cell motility and resistance to antibacterial agents (Farris *et al.*, 1998; Duo *et al.*, 2008). Furthermore, it was shown to regulate cell surface and virulence associated components such as flagellar, the *espC* gene from the EspC pathogenicity island and type III secretion systems indicating a high hierarchical role of *bipA* to control virulence associated mechanisms in pathogenic *E. coli* (Grant *et al.*, 2003). Another study in *Escherichia coli* showed that the induction of *bipA* is growth-phase dependent and that *bipA* is required for the expression of Fis, a highly ranked transcriptional modulator in *E. coli* (Owens *et al.*, 2004). Pfennig & Flower (2001) demonstrated that *bipA* was required for growth at low temperatures. In *Sinorhizobium meliloti typA* mutants (*typA* is the orthologue to *bipA* in *E. coli*) were sensitive to low temperature and low pH as well to exposure to sodium dodecyl sulfate (SDS). Additionally, *typA* was required for efficient symbiosis in some host plants (Kiss *et al.*, 2004). Krishnan & Flower (2008) verified that ribosomal structure and function is dependent on BipA. Thus, they hypothesized in accordance with the previously identified *bipA*-dependent phenotypes that BipA is a novel regulatory protein responsible for efficient translation of target genes by direct interactions with the ribosomes.

1.3 Gene Expression Analysis

The genome-wide functional characterization of microorganisms has rapidly evolved since the availability of the first complete microbial genome sequence (Fleischmann *et al.*, 1995). The field of functional genomics focuses on the whole-genome analysis by combining different approaches to study the genes of an organism, their resulting proteins and their functional role in regulatory and biochemical processes: i) the analysis of transposon mutant libraries allows the identification of essential genes required under certain growth conditions (Hensel *et al.*, 1995), ii) the identification of genes with related functions can be assessed by systematic pairwise deletion or inhibition of genes, iii) gene expression profiling by microarrays enables the targeted expression analysis of every single gene of a genome under defined conditions (Brown & Botstein, 1999), iv) the proteomics approach facilitates identification of global protein expression of a cell under various conditions to reveal their physicochemical properties and to describe their function and regulation (Griffin & Aebersold, 2001) as well as to analyze protein-protein interactions (Fromont-Racine *et al.*, 2000, 2002). Two-dimensional gel electrophoresis with subsequent mass spectrometry analysis was used as standard method for the identification of target proteins and the characterization of the proteome under different conditions (Henzel *et al.*, 1993). A new method, peptide mass fingerprinting by MALDI-MS and subsequent sequencing by tandem mass spectrometry, was recently developed and helps to overcome limitations by two-dimensional gel electrophoresis (Thiede *et al.*, 2005).

Among the different approaches of functional genomics, the field of transcriptome analysis was advanced by the introduction of deep sequencing technologies providing a potential tool for expression analysis without the limitations of hybridization to predefined target sequences, as is the case with microarrays (Wang *et al.*, 2009).
Since the 1990s when the first bacterial genomes were published, *Haemophilus influenzae* (Fleischmann *et al.*, 1995), *E. coli* (Blattner *et al.*, 1997) and *Mycobacterium tuberculosis* (Cole *et al.*, 1998), sequencing technologies have rapidly improved resulting in the vast amount of available sequence data today. The genome sequences opened the way to the "omics" field including transcriptomics. Microarrays were the optimal technology for gene expression analysis for some time, but are now having to compete with the deep sequencing technologies. By simultaneous sequencing of millions of fragments based on reversible terminator chemistry, the resulting millions of short reads from RNA/cDNA samples can be mapped to the reference genome and are thus also suitable for transcriptome analysis.

Many reviews have already addressed the questions of the comparability of microarrays and deep sequencing approaches (Metzker, 2005; Hall, 2007; Tucker *et al.*, 2009; Wang *et al.*, 2009; Willenbrock *et al.*, 2009). Beside the ongoing discussion about reproducibility and comparability of results among different laboratories investigating the same biological questions ('t Hoen *et al.*, 2008; Pedotti *et al.*, 2008), the deep sequencing technologies can overcome some of the limitations of microarrays:

- Reliance on known genome sequence
- Cross-hybridization resulting in high levels of background noise
- Background and saturation of signals resulting in limited dynamic range of detection
- Complicated normalization methods for comparable expression levels

Table 1.1   Comparing massively parallel sequencing technologies (Tucker *et al.*, 2009).

	Sequencing chemistry	Amplification approach	Read length	Run time and throughput	Raw accuracy	Cost
Illumina	PCR-based sequencing by synthesis	Bridge PCR	75+ bp	17 Gb in 7 days	98.5%	$6/Mb
SOLiD	Ligation-based	Emulsion PCR	50 bp	10-15 Gb in 3-7 days	99.94%	$5.8/Mb
454	Pyrosequencing	Emulsion PCR	400 bp	400-600 Mb in 10 hrs	99%	$84.4/Mb
Helicos	polymerase-based	none	30-35 bp	21-28 Gb in 8 days	99%	Not available

However, high-throughput sequencing methods such as Roche 454 and Illumina genome Analyzer still rely on PCR-based amplification steps that can introduce technical bias, but that can be overcome in future for deep RNA sequencing technologies by amplification-free sequencing-library preparation. Furthermore, this new method also improves mapping and assembly of GC-biased genomes (Kozarewa *et al.*, 2009). One disadvantage of high-throughput sequencing are currently still the higher costs compared to microarray platforms, but with further development of these technologies, costs will decrease as already indicated by the comparison of 454 sequencing and the new deep RNA sequencing technologies (see Table 1.1).

Though the short read length (36 – 75 bp) provided by the new generation sequencing technologies make *de novo* assembly of novel genomes difficult, this is not a limitation in transcriptome analysis. In fact, transcriptome analysis using deep RNA sequencing technologies has revolutionized gene expression profiling by the strand-specific identification of novel sRNAs, antisense activity within coding regions and transcriptional start sites and operons and helps to reannotate genomes by detecting new protein coding reading frames (Oliver *et al.*, 2009; Albrecht *et al.*, 2010; Filiatraut *et al.*, 2010; Sharma *et al.*, 2010; Sorek & Cossart, 2010).

The first bacterial transcriptome derived from deep RNA sequencing was described by Oliver *et al.* (2009). They analyzed the transcriptome of *Listeria monocytogenes* upon stationary phase stress. By comparing the transcriptome of the *L. monocytogenes* wild type and isogenic σ^B mutant, they revealed many σ^B-dependent genes contributing to the stress response. In total, they identified 96 genes with σ^B-dependent expression and 67 sRNAs that were transcribed in stationary phase. In combination with a dynamically trained Hidden Markov Model, they identified 65 σ^B promoter regions upstream of the identified σ^B-dependent genes.

Sharma *et al.* (2010) recently published the primary transcriptome of *Helicobacter pylori*. By using a novel approach, called dRNA-Seq, which is selective for the 5' end of the transcript, they discovered several hundred transcription start sites within predicted operons, as well as antisense within annotated genes. Furthermore, they revealed potential regulators of sense and antisense-encoded mRNAs, and a high number of previously undetected sRNAs. Overall, they confirmed 60 transcripts to encode sRNAs by Northern Blot.

A third study on bacterial transcriptomes was published on *Pseudomonas syringae* (Filiatraut *et al.*, 2010). They used a strand-specific method for one enriched RNA sample. In combination with proteomics and bioinformatic analysis, they were able to identify transcription in so far unannotated regions and transcription inconsistent with the current annotation, leading to a partial reannotation of the *Pseudomonas syringae* genome. Furthermore, they validated a few transcriptional start sites and found potential RpoN-dependent promoter sequences upstream of sRNAs, suggesting a role of these sRNAs in RpoN regulation.

These studies on bacterial transcriptomes using the Illumina Genome Analyzer clearly demonstrate the advantages of deep RNA sequencing in detecting novel transcripts such as small regulatory RNAs, identifying transcription start sites, distinguishing between sense and antisense expression by strand-specific amplification and in helping to reannotate the bacterial genome by detecting novel protein coding open reading frames or transcription inconsistent to the existing annotation.

1.4 Systems Biology

Systems biology is an emerging field in the biological sciences that is derived from the development of high-throughput methods in the genomics field. It is an approach for the comprehensive quantitative analysis of a biological system to understand and elucidate the functionality and interaction of the systems' compartments as a whole. To achieve this, an interdisciplinary collaboration of scientists from different research fields is essential for an integration of experimentally and computationally derived data.

According to Aitchison & Galitski (2003) three major concepts make systems biology unique in the field of bioscience research:

- Definition of all relevant elements of a system, and the quantitative analysis of its interactions in normal and perturbed state by high-throughput data generation.
- Integration of multiple data types for a complex understanding of the systems properties on the different hierarchical levels.
- Formulation of quantitative system models for the computational analysis and visualization of biological processes to generate new hypotheses of the systems behavior to certain perturbations that can then again be experimentally tested.

The European transnational funding and research initiative on "Systems Biology of Microorganisms (SysMO)" started in May 2007 with the goal to describe dynamic processes in unicellular microorganisms and to visualize them in computer-driven mathematical models. Within this initiative, eleven projects have been funded working on various microorganisms such as *Bacillus subtilis*, *Clostridium acetobutylicum*, Lactic Acid Bacteria, *Saccharomyces cerevisiae*, *Streptomyces coelicolor*, *Sulfolobus solfataricus*, *Pseudomonas putida* and *Pseudomonas fluorescens* with the aim of advancing biomedical and biotechnological research. PSYSMO is one out of eleven projects of the research initiative with focus on the metabolically versatile soil bacterium *Pseudomonas putida*. This project "Systems analysis of biotech induced stresses: towards a quantum increase in process performance in the cell factory *Pseudomonas putida*" aimed to develop a systems understanding for its potential in biotechnological applications. The long-term goal of this project is, by linking systems and synthetic biology, to reprogram the cell for the directed production of industrially important metabolites and to predict the stress response to process-induced perturbations being either internal by hyperproduction of metabolites or external due to solvent, solute or temperature stress. This involves the following points:

- **Establishment of a systems biology framework** for the study of *P. putida* by combining modelling and experimentation in all stages.
- **Identification of critical structural and regulatory components** in *P. putida*, and detailed characterization of the subset of network connections and nodes linking responses to process-induced stresses and cell factory productivity.
- **Development and experimental validation** of:
    o Genome-scale, constraint-based models and thereby defining their global metabolic space, network properties, optimality lines and flux distribution potential.
    o Models for the inference of regulatory networks from high-throughput data.
    o Detailed models of specific biotransformation circuits of interest, thereby generating a rigorous description of their dynamic behaviour, regulatory circuitry and functionality.
- **Elucidation, quantification and prediction** of the interplay of cellular activities of interest (e.g. production of metabolites of aromatic hydrocarbon degradation, enzyme hyper-secretion, PHA production) with host genome, with the help of the genome-scale models and electronic-like descriptions developed.
- **Production of *in-silico P. putida*** (partial) blueprints and new conceptual models that
    o Account for essential features of this bacterium under a range of conditions.
    o Connect signal inputs to metabolic and regulatory properties of the networks and thereby to their biological outputs.
- **Use of the predictive models** developed and validated above for the targeted improvement of model bioprocesses, which will serve as proof-of-principle.

1.5 Objectives of the Present Study

The present study was incorporated into the European research consortium PSYSMO working on *Pseudomonas putida* with a systems biology approach to understand its potential for biotechnological applications. Generally, biotechnical applications derive from the organisms' capacity to withstand environmental stresses to which it is exposed in its natural habitat or to catabolize unfavourable carbon sources. In this context, this study focuses on the response to two stress conditions, benzoate and cold stress, to monitor environmental stresses under defined conditions.

The aromatic acid benzoate represents a good carbon source to monitor the stress response to high concentrations of aromatic carbons. The degradation of benzoate to catechol that funnels via the ß-ketoadipate pathway into the citric acid cycle is well characterized and thus is suitable for studying perturbations caused by high concentrations of the aromatic compounds.

Here, the transcriptome profile of a *P. putida* KT2440 chemostat culture in response to a pulse of benzoate resulting in a 45 mM increase in concentration is analyzed over a time period of three hours. Monitoring the expression pattern over a long time period should reveal the genes involved in the direct stress response regulated directly after the pulse, beside those that are known for the direct conversion of benzoate via the ß-ketoadipate pathway and to find genes responsible for the adaptation process and switch to normal cell state once the cells started to reproduce again.

The cold shock response of *P. putida* KT2440 to a temperature shift from 30°C to 10°C is analyzed mainly on the transcriptional level. In a systems biology approach, the very same samples were used for proteome and metabolome analysis to complete the transcriptome profile and to examine the three "omics" for their comparability. In addition, five transposon mutants, *cbrA*, *cbrB*, *pcnB*, *vacB* and *bipA*, that had been identified in a previous study to be cold sensitive (Reva *et al.*, 2006), are analyzed for the transcriptome and metabolome profile in response to cold shock. The comparison of cold sensitive mutants and the wild type should reveal new mechanisms in *P. putida* KT2440 for coping with a sudden decrease in temperature.

A promising method of transcriptome analysis is the deep RNA sequencing technology. Therefore, we used wild type samples from the cold shock experiment to compare two different DNA microarrays available for *P. putida* (purchased from the companies Progenika and Affymetrix), and deep RNA sequencing using the Illumina Genome Analyzer for comparison of the gene expression profiles derived from the three transcriptome platforms. Furthermore, the key players in the stress response of *P. putida* KT2440 upon cold shock should be determined.

Recent studies have demonstrated the high abundance of coding frames such as for small RNAs and proteins in the intergenic regions of prokaryotic genomes (Hemm *et al.*, 2008, 2010; Oliver *et al.*, 2009; Filiatraut *et al.*, 2010; Sharma *et al.*, 2010). Thus, the deep RNA sequencing data were furthermore used for the expression analysis of intergenic regions, since they have been hardly analyzed so far and knowledge about their role and functions is scarce in *P. putida*.

2 MATERIALS AND METHODS

2.1 Materials

2.1.1 Equipment
The equipment used during this study is recorded in association with the methods described in this chapter.

2.1.2 Consumables

2.1.2.1 *Pseudomonas putida* Genome Oligonucleotide Array
The *P. putida* Genome Oligonucleotide Array (Progenika Biopharma, S.A, Derio, Spain) was developed in collaboration with the company Progenika Biopharma and several Spanish scientists working in the field of *P. putida* (Dr. Fernando Rojo, Madrid; Dr. Juan Luis Ramos, Granada; Dr. Eduardo Díaz, Madrid; Dr. Victor de Lorenzo, Madrid and Dr. Eduardo Santero, Seville).

It is a two-dye gene expression array, whereby the comparison of the binding efficiencies of two samples to one array provides an insight into gene expression changes in a single experiment. In this protocol the two fluorescent cyanine dyes cy3 and cy5 are used for indirect labelling by coupling to aa-dUTP, an amine-modified nucleotide incorporated into cDNA during reverse transcription (see 2.7.4.1 and 2.7.5.1).

The array is designed as listed below:

- γ-aminosilane treated glass slides
- single-stranded oligonucleotides in repeating spots
- 50mer oligonucleotides
- One oligonucleotide representative for one ORF
- 5539 *P. putida* KT2440 ORFs
- Homogeneity control with 2 ORFs (20 replicates each) distributed over entire array
- Negative control with DMSO (50%), 406 spots, for background hybridization correction distributed over entire array

2.1.2.2 Affymetrix microarray

The Affymetrix high-density oligonucleotide microarray was based on the annotated genome of *P. putida* KT2440 (NC_002947.3) and was designed with a pair-wise configuration of 13 perfect match (PM) and mismatch (MM) 25mer oligonucleotides per probe set. In total 8047 probe sets were spotted that represent 5330 annotated ORFs, 207 ORFs not present on the *Pseudomonas* homepage (www.pseudomonas.com), 22 genes encoded on the TOL plasmid (*xyl* cluster), 2443 intergenic regions and 45 control sequences from other organisms not in the genus *Pseudomonas*. Preparation of RNA, cDNA and cDNA fragmentation and biotin-labelling are described in chapters 2.7.2., 2.7.4. and 2.7.5.

2.1.2.3 Consumables

A selection of further consumables used in this study is given below:

Consumable	Manufacturer
CryoTubeTM Vials	NUNC
Electroporation cuvettes 1 mm	BioRad
Filter Celluloseester HA, 0.45 µM pore size	Millipore
Hybond N+ Nylon Membrane	Amersham Biosciences
MicroCons YM-10	Millipore
Qiaprep Spin Miniprep Kit	Qiagen
QiaQuick PCR Purification Kit	Qiagen
RNeasy Mini Kit	Qiagen
Whatman paper	Schleicher & Schüll
X-ray film X-Omat AR	Qiagen
96 well plates	Greiner

2.1.2 Chemicals

Chemical	Manufacturer
Agarose	Eurogentec
Antifoam Struktol SB2121	Schill & Seilacher
Biotin, Streptavidin (anti-streptavidin (goat), biotylilated)	Vector Laboritoiries
BSA	New England Biolabs
CDP-Star	Tropix
Coomassie Brilliant Blue R250	Serva
Cyanine-3, -5	GE Healthcare
λ-DNA, BstEII digested	NEB
50bp-, 100bp-ladder	Fermentas
Ethanol	J.T. Baker
Formaldehyde	Merck
GeneChip DNA Labelling	Affymetrix
Gentamicin	Serva
Glycerol	AppliChem
β-Mercaptoethanol	Merck
Neutravidin	ThermoScientific
Oligonucleotides	MWG
10x One-Phor-All Buffer	Amersham
Phenolrot	Merck
RNAprotect Bacteria Reagent	Qiagen
Rnase Inhibitor	Ambion
R-Phycoerythrin Streptavidin (SAPE)	DIANOVA
Sephadex G-50	Pharmacia Fine Chemicals
Sucrose loading dye	Amresco
SYBR Gold	Invitrogen

All chemicals not listed were either purchased from Fluka, Roche, Roth or Sigma-Aldrich.

2.1.3 Enzymes

A list of enzymes with their corresponding buffer system is listed below:

Enzyme	Manufacturer
Alkaline phosphatase, calf intestine (CIP)	NEB
Anti-digoxigenine alkaline phosphatase	Roche
DNase I	Qiagen
FailSafe^{TM} PCR Premix Selection Kit	Epicentre Technologies
Goldstar-Taq-Polymerase	Eurogentec
Invitek-Taq-Polymerase	Invitek
Klenow-Poylmerase	Boehringer
Restriction endonucleases + buffer systems Acc65I, BamHI, EcoRI, HindIII, NlaIII, NsiI, PstI, Sau3AI, SphI, XbaI	NEB
RNase A (10 mg/ml)	Qiagen
SuperScriptII reverse transcriptase	Invitrogen
Terminal transferase	Promega
T4 DNA ligase	NEB
T4 PN kinase	NEB

2.2 Media and Solutions

Media and consumables were sterilized by autoclaving at 121 °C for at least 30 minutes, unless otherwise specified. The reagents were of high purity ("pro analysis") and purchased from Fluka, Merck, Roche, Roth or Sigma-Aldrich.

2.2.1 Media

Luria-Bertani Medium (LB Medium)

Tryptone	15.0 g/L
Yeast Extract	5.0 g/L
NaCl	10.0 g/L
pH 7.0	

LB-Gm:	LB medium/ agar with 30 µg/mL Gentamicin
LB-Amp:	LB medium/ agar with 100 µg/mL Ampicillin
LB-Car:	LB medium/ agar with 1000 µg/mL Carbenicillin

LB Agar: LB medium was solidified by adding 20 g/L agar and subsequent autoclaving.

Glycerol Medium

For long-term storage of bacterial strains LB medium was supplemented with glycerol to a final concentration of 15%.

ABC Minimal Medium

Na_2HPO_4	6.0 g/L
KH_2PO_4	3.0 g/L
NaCl	3.0 g/L
$(NH_4)_2SO_4$	2.0 g/L
pH 7.0	

Medium was supplemented with 15 mM or 45 mM sodium benzoate for phenotypical verification growth experiments.

M9-Minimal Medium (5x)

Na_2HPO_4	33.9 g/L
KH_2PO_4	15.0 g/L
NaCl	2.5 g/L
NH_4Cl	5.0 g/L
pH 6.8	

Minimal Medium for growth experiments

	stock concentration	end concentration
M9-Minimal medium	5x	1x
Succinate	1 M	15 mM
MgSO4	1 M	2 mM
$CaCl_2$	0.1 M	0.1 mM
$FeSO_4$ x 7 H_2O	50 mM	0.01 mM
Trace elements	1000x	1x

Stock solutions were adjusted to a pH of 6.8, filter sterilised through a 0.2 µm nitrocellulose filter and stored aseptically prior to use. $FeSO_4$-solution was always prepared fresh. Solutions were mixed immediately before usage.

2.2.2 Solutions

Antibody-Solution

Antibody-solution was made by adding 10 µL Anti-Digoxigenin AP F_{ab} (150 U/200 µL) to 50 mL Buffer II (1:5000 dilution)

Antibody solution for Affymetrix microarray hybridization

2x staining buffer	315 µL
DEPC-treated H2O	279.7 µL
BSA (50 mg/mL)	25.2 µL
Goat IgG (10 mg/mL)	6.3 µL
Biotin / streptavidin	3.8 µL

Blocking buffer

SSC	5% (v/v)
SDS	0.1% (w/v)
BSA	1% (v/v)

Blotting Buffer

NaOH	0.4 M

Blot washing buffer

sodium phosphate   50 mM
pH 6.5

Buffer I

Tris/HCl   100 mM
NaCl   150 mM
pH 7.5

Buffer II

Buffer II was freshly made from Buffer I by adding 0.5% blocking reagent (Roche). The solution was stirred on a heater to ensure solubility.

Buffer III

Tris/HCl   100 mM
NaCl   100 mM
$MgCl_2$ 50 mM
pH 9.5

CDP Star

A 12.5 mM stock solution of CDP Star was stored at 4°C and diluted 1:500 in Buffer III prior to use.

Coomassie solution

Coomassie Brilliant Blue R250   0.05%
Methanol   50%
Acetic acid   10%

Denhardt's solution (50x)

BSA   1% (w/v)
Ficoll   1% (w/v)
PVP   1% (w/v)

Solution was sterile filtrated (Ø 0.2 µm) and stored at 4°C.

Elution-buffer

Potassium Phosphate Buffer   5 mM
pH 8.5

Hybridization buffer

Formamide (deionised)	50% (v/v)
Denhardt´s solution	5% (v/v)
SDS	1% (w/v)
SSC	3% (v/v)
dextransulphate	5% (w/v)

Hybridization buffer (2x) for Affymetrix mmicroarrays

12x MES	8.3 mL
NaCl (5 M)	17.7 mL
EDTA (0.5 M)	4 mL
Tween 20 (10%)	0.1 mL
ddH2O	19.9 mL

Killing buffer

Tris-HCl	20 mM
$MgCl_2$	5 mM
NaN_3	20 mM
pH 7.5	

Loading Buffer (6X)

Ficoll 400	15% (v/v)
Bromophenol Blue	0.25% (w/v)
Xylene cyanol	0.25% (w/v)
EDTA	0.5 M
pH 8.0	

Lysis Buffer

Tris-acetate	40 mM
Sodium acetate	20 mM
EDTA	1 mM
SDS	1% (w/v)
pH 7.8	

12x MES stock buffer

MES free acid monohydrate	17.6 g
MES sodium salt	48.33 g
ddH2O	add to 250 mL
pH 6.5 – 6.7	

The solution was sterile filtrated and stores at 4°C in the dark.

MOPS buffer (10x)

MOPS	200 mM
Sodium acetate	100 mM
EDTA	10 mM
pH 7.0	

Neutralization Buffer

NaH_2PO_4	40 mM
pH 6.5	

Potassium Phosphate Buffer

KH_2PO_4	10 mM
K_2HPO_4	10 mM

Prehybridisation Buffer

SDS	7% (w/v)
NaH_2PO_4	0.5 M
EDTA	1 mM
Blocking Reagent	0.5%
pH 7.2	

RNA loading buffer

Glycerol	50% (v/v)
EDTA	1 mM
Bromphenol blue	0.25 % (w/v)
pH 6.0	

SAPE stain solution

2x staining buffer	630 µL
DEPC-treated H_2O	567 µL
BSA (50 mg/mL)	50.4 µL
SAPE (0.5 mg/mL)	12.6 µL

Solution I

Tris/HCl	50 mM
EDTA	10 mM
DNase free RNase A	100 µg/mL
pH 8.0	

Solution II

NaOH	0.4 M
SDS	1% (w/v)

Solution III

Potassium acetate	3 M
Acetic acid	2 M

20x SSPE solution

Sodium chloride	3 M
Sodium phosphate	200 mM
EDTA	20 mM
pH 7.7	

2x Staining buffer for Fluidic station (Affymetrix)

12x MES	41.65 mL
NaCl (5 M)	92.5 mL
Tween 20 (10%)	2.5 mL
ddH$_2$O	112.8 mL

Standard saline citrate solution (20%) (SSC)

Sodium chloride	175.3 g
Sodium citrate	88.2 g
pH 7.0	

Stripping buffer

NaOH	0.2 M
SDS	0.1% (w/v)

TBE-Buffer (10X)

Tris/HCl	0.9 M
Boric Acid	0.9 M
EDTA	0.02 M
pH 8.3 - 8.5	

TE Buffer

Tris/HCl	10 mM
EDTA	1 mM
pH 8.0	

Trace elements (1000x)

$ZnSO_4 \times 7\ H_2O$	0.148 g/L
$MnCl_2 \times 4\ H_2O$	0.1 g/L
$CoSO_4 \times 7\ H_2O$	0.236 g/L
$NiCl_2 \times 6\ H_2O$	0.1 g/L
$CuCl_2 \times 2\ H_2O$	0.02 g/L
$NaMoO_4 \times 2\ H_2O$	0.05 g
HCl (36%)	1 mL

1x UT solution

Urea	8 M
Thiourea	2 M

Washing Buffer for Southern Blot

NaH_2PO_4	40 mM
SDS	1% (w/v)
1 mM EDTA	
pH 7.2	

Washing-buffer for cDNA labelling

Potassium phosphate buffer	4 mM
Ethanol	80% (v/v)
pH 8.0	

Washing buffers for microarray hybridization (Progenika)

WB1	SSC 2%, SDS 0.1%
WB2	SSC 1%
WB3	SSC 0.2%
WB4	SSC 0.1%

Washing buffers for Fluidic Station (Affymetrix)

Buffer A

20x SSPE	150 mL
Tween 20 (10%)	0.5 mL
ddH_2O	add to 500 mL

Buffer B

12x MES	41.65 mL
NaCl (5 M)	2.6 mL
Tween 20 (10%)	0.5 mL
ddH_2O	add to 500 mL

2.3 Biological Materials

The *P. putida* KT2440 mini Tn5 plasposon mutant library was generated by Christian Weinel (2003). Insertion point and flanking genes were determined either by plasmid rescue or Y-linker method. Plasposon mutants used in this study were already described for their stress sensitive phenotype (Reva *et al.*, 2006).

2.3.1 Strains

Table 2.1 Overview of strains used in this study for either cold or benzoate stress experiments.

Strain	Genotype	Reference
Pseudomonas putida KT2440 (DSM 6125)	hsdR1, hsdM$^+$, Ben$^+$	Bagdasarian et al., 1981
#cysM	pTnMod-OGm::KT*cysM*:92*	Reva et al., 2006
#fepA	pTnMod-OGm::KT*fepA*:1831	
#nuoL	pTnMod-OGm::KT*nuoL*:840	
#kgdA	pTnMod-OGm::KT*kgdA*:2337	
#PP4646	pTnMod-OGm::KTPP4646:762	
#cbrA	pTnMod-OGm::KT*cbrA*:263	
#cbrB	pTnMod-OGm::KT*cbrB*:948	
#pcnB	pTnMod-OGm::KT*pcnB*:731	
#omlA	pTnMod-OGm::KT*omlA*:394	
#vacB	pTnMod-OGm::KT*vacB*:970	
#bipA (358)	pTnMod-OGm::KT*bipA*:358	
#pstB	pTnMod-OGm::KT*pstB*:561	
#pstA	pTnMod-OGm::KT*pstA*:262	
#pstC	pTnMod-OGm::KT*pstC*:1310	
#pstS	pTnMod-OGm::KT*pstS*:596	
E. coli DH5α	F$^-$ ϕ80*lacZ*ΔM15 Δ(*lacYZA-argF*)U169 *recA1 endA1 hsdR17* (r_K^- m_K^+) *supE44 thi-1 gyrA relA1* (Nal^r)	Ausubel et al., 1987

* Number indicates insertion site within the gene.

2.3.2 Plasmids

Table 2.2 Overview of plasmids used in this study.

Plasmid	Genotype	Reference
pUCP20	*Escherichia – Pseudomonas* shuttle vector; Ap^r	Garrity-Ryan et al., 2000
pUCP20::KT*cysM*	pUCP20 plasmid carrying the XbaI/HindIII the PCR product bearing the *cysM* gene	This study
pUCP20::KT*kgdA*	pUCP20 plasmid carrying the XbaI/HindIII the PCR product bearing the *kgdA* gene	
pUCP20::KTPP4646	pUCP20 plasmid carrying the XbaI/HindIII the PCR product bearing the PP4646 gene	
pUCP20::KT*omlA*	pUCP20 plasmid carrying the SphII/HindIII the PCR product bearing the *omlA* gene	

2.3.3 Oligonucleotides

Table 2.3 Summary of oligonucleotides used in this study.

Primer in 5' – 3' direction				Reference
Verification of plasposon mutants				
F_0063	CTGCTACGCCTTCATGCAGG	R_0063	TGGTGCGATCATCCGTGACG	This study
F_0368	CGACAGCCGTATCTCCATGC	R_0368	AGGCGATGTACACCGTCAGC	
F_0816	GGTGATCGCTCTGCTTGTCG	R_0816	GTCCACACCCATCATCACGC	
F_16541	GCCATGACCTTGCAGTACCC	R_1654	GCTGTTGTAGTGGGCGATCG	
F_1910	GCACTCCAGTAGTTGCTCGC	R_1910	CCTCGTCCTTGCTCGATTCG	
F_2242	CGACCACCACGAAACCTTCG	R_2242	ACTTGTACCCGGCACTCACG	
F_3820	TGGCCTGACGAACAGGAAGC	R_3820	GTTCTCTCAGGAAGCGCTCG	
F_3951	CAAGGTGGTCTGCTCGTTCC	R_3951	CTTGACCGACACCTGGAAGG	
F_4129	CATCCTGGTACATGCGTGGC	R_4129	AGGCTTCCCAGAGGATCTCG	
F_4185	GCTGATCGTCTGCATCACCG	R_4185	GGGTAAGGGGAAATGCTGCC	
F_4186	CGAGGGCGTTGAGGTTAACG	R_4168	CCATCGGCCTTGATCACCAG	
F_4189	CAGTGGTGAGCACAAGTGGG	R_4189	CAGATCGTCCTCAGGGAACG	
F_4646	GATCCTGCAGGACTTCGAGG	R_4646	CGGTCGGTATCATCATCGGC	
F_4695	GTGGATTCCCGCTAGCATCC	R_4695	GGTACATGTGCGGCATGACG	
F_4696	GCCTACCGACTCCAATGTGC	R_4696	GCTGGTAGCGTTGCTCTTGC	
F_4697	CCACCGTTCGTCGCATAACG	R_4697	CAAGCTCGCGGATCATCTCG	
F_4731	CGAAGTTGTGGCGAACCACC	R_4731	GCGATACACACATGCGTGCC	
F_4880	TGCCAAGATCGGCCAGTTCG	R_4880	CTCCTCGCCGAAGATGATGC	
F_48801	GAACAAAGCCGGCCAGATGG	R_48801	GTTGCTCCAGGGTGTTCTCG	
F_4941	GTTGCTGCTGGACGGTAAGC	R_4941	TGCTCGGCCTCTATCTCTCC	
F_4982	CAGCATCATCCTGCACTGGG	R_4982	GGTAGCCTTGAACGTCACCG	
F_5044	GCTGTACCAGTCGATCGTCG	R_5044	CTCGACCGAAACCAGTACGC	
F_50441	CAAACTCCTGCGTCAGTCCG	R_50441	AGGTCATCGAAACGGTCGGC	
F_5322	GCCTCACCTCGTTATTCGCC	R_5322	CTTCCATGGTCAGGTAGCCG	
F_5326	TTCCCACACCCATGGCATCG	R_5326	CTTCGGTCTGCTTCTTCGCC	
F_5327	CCACTATCGCGTGCTGTTCC	R_5327	GGGTCATCATCACGGTACCG	
F_5328	TCCGTACCTTCCAGATGGGC	R_5328	CAGCACGATCATGGTCTCGC	
F_5329	TCCCGACCTACACCAAGACC	R_5329	GCAGCGGAATGTAGCCATCC	
Genta-f	AGCCGATCTCGGCTTGAACG	Genta-r	CAAGCAGCAAGCGCGTTACG	
trans1	CTTGGTCGAAGGCAGCAAGC			
Complementation				
F_1654:XbaI	CGTGTCTAGATGCCTTGGTGATCCTTGCACG			This study
R_1654:HindIII	CGAGGAAGCTTGGTCGTAGTAGCGACATGGC			
F_4189:XbaI	CGTGTCTAGATGACCCGTTCAGCGTATTCCGC			
R_HindIII	CGAGGAAGCTTGCTCGTCACGCTTGACGGCTTC			
F_4646:XbaI	CGTGTCTAGATGTAGAGGTACAGCCTTGGCG			
R_4731:HindIII	CGAGGAAGCTTTGCCTTCGGCAGAAGAGACG			
F_4731:SphI	CGTGCGGCATGCTCCACGTTCACCATGTGGTCG			
R_4731:HindIII	CGAGGAAGCTTTGGCATGAGCCAGCACTTCG			
Y-linker				
Y-primer	CTGCTCGAGCTCAAGCTTCG			Kwon & Ricke, 2000
Tn2	TGCGTTCGGTCAAGGTTCTGG			
Y1:NlaIII	TTTCTGCTCGAGCTCAAGCTTCGAACGATGTACGGGGACACATG			
Y2:NlaIII	TGTCCCCGTACATCGTTCGAACTACTCGTACCATCCACAT			
Y1:Sau3AI	TTTCTGCTCGAGCTCAAGCTTCGAACGATGTACGGGGACA			
Y2:Sau3AI	GATCTGTCCCCGTACATCGTTCGAACTACTCGTACCATCCACAT			
RT/PCR				
F_ttgR	AGCCTGCACGAGACCCATGACC	R_ttgR	CAAGCGCAGCATATCCAGCCCG	This study
F_algU	TTCTCGGGTTGATCGTGCGG	R_algU	ATTCGAGAGCGCACGGTACC	
F_benR	TGGCTTCGTGAACCTGCTGG	R_benR	CAACTCGCCAAACTGCTGCC	
F_benF	AGATGAACGCGATGCTCGGC	R_benF	ATAGGTCGGCACCTCGTTGC	
F_catB	CTCAAGCTGGACAAGCTGGC	R_catB	CTGCTCGATCAGGTCGATGC	

2.4 BioFlo 110 Modular Benchtop Fermenter

Growth experiments for comparative analysis of transcriptome, metabolome and proteome in *P. putida* wt and respective mutants were performed in a bioreactor with 1.5 L operating volume.

The BioFlo 110 fermenter (New Brunswick Scientific Co., Inc., Edison, NJ, USA) included a vessel with an external heat blanket and a headplate for variable port usage. pH and dO_2-probe were provided by Mettler-Toledo GmbH (Gießen). The process was controlled by the Primary Control Unit (PCU), including agitation, aeration, pH, dO2 and antifoam. The *BioCommand* software was used for documentation and visualization of the fermentation process (both New Brunswick Scientific Co., Inc., Edison, NJ, USA).

Media was maintained to 30°C, agitated with 600 rpm and streamed with pressure air (~ 21% oxygen) with an air flow of 1.5 L/min. After dO2-probe calibration, oxygen saturation was set to 100%. Agitation, aeration and pH were kept constant during fermentation. pH was adjusted automatically to 6.8 ± 0.1 with 5 % H_3PO_4 or 1 M NaOH. Antifoam was provided with the medium (0.025%).

The vessel containing pure M9 medium with 0.025% antifoam and all required probes were autoclaved at 121°C for 30 min ahead of fermentation, carbon source and other components of the media were aseptically added afterwards.

To check for contamination, prior to inoculation, 1 mL media were collected and screened for bacterial growth on LB agar plates.

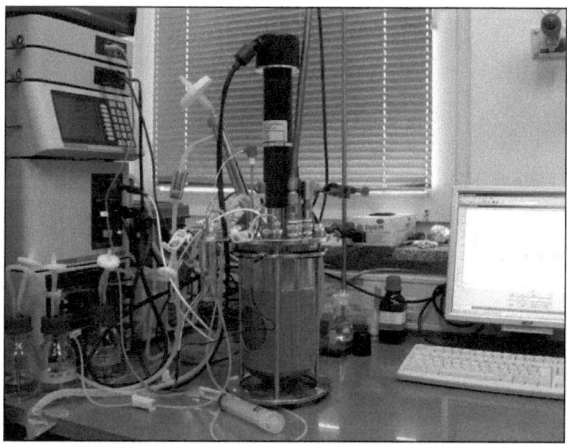

Fig. 2.1 BioFlo 110 Modular Benchtop Fermenter.

2.5 Microbiological Methods

2.5.1 Bacterial growth conditions

Bacterial cultures of *P. putida* or *E. coli* DH5α were inoculated from glycerol-stocks in LB medium and grown at 30°C or 37°C (250 rpm) overnight (12-16 hours), respectively.
With pUCP20 transfected *P. putida* or *E. coli* DH5α strains were cultured in the presence of 1000 µg/mL carbenicillin or 100 µg/mL ampicillin, respectively.

2.5.2 Determination of cell density

Optical density (OD) of growing cultures was measured spectrophotometrically at 600 nm (Spectrophotometer U3000, Hitachi).
Correlation of OD in LB and colony forming units is as follows:

P. putida $OD_{600\,nm} = 1.0 \rightarrow 1 \times 10^9$ cfu/mL
E. coli $OD_{600\,nm} = 1.0 \rightarrow 0.8 \times 10^9$ cfu/mL

2.5.3 Determination of colony forming units

P. putida cultures were grown in respective media until stationary phase was reached. During growth 1 mL aliquots were taken and OD was determined. According to the OD a serial dilution in 0.9% NaCl was produced to obtain 30 to 300 cfu/mL. 100 µL of two dilutions was spread in duplicates onto LB agar plates and incubated at 30°C for 20 to 24 hours. Single colonies were counted and cfu/mL calculated.

2.5.4 Determination of cell dry weight

Cell dry weight in correlation to optical density was determined using a 1225 sampling Manifold (Millipore). This device includes 12 sampling ports that hold a volume up to 15 mL. A filter, 0.45 µM pore size, is placed on the filter gasket to retain bacterial material. The sampling manifold can be connected to a vacuum pump and samples can be easily filtrated.
A bacterial culture was grown in respective media, here M9 media supplemented with 15 mM succinate as reference media, to late exponential phase and a serial dilution was prepared from which the OD was measured. From each dilution 10 mL in triplicates were loaded on the filter

devices and drawn by vacuum. The filter with retarded cell mass was then incubated at 65°C for 2 days until all liquid was evaporated. The mean difference in weight of empty and saturated filter gave the cell dry weight for each optical density measured. A linear correlation of optical density and cell dry weight allowed the calculation of sample volume needed for metabolome analysis.

2.5.5 Maintenance of bacterial cultures

For long-term storage bacterial cultures were maintained in glycerol medium and stored in cryotubes at -80°C.

2.5.6 Growth and purification of plasposon mutants

The *P. putida* plasposon mutant library was stored in 96 well plates at -80°C.

Candidate mutants for this study were grown overnight in LB containing 30 µg/mL gentamicin at 30°C (250 rpm). Overnight cultures were diluted in LB and spread onto LB plates containing again 30 µg/mL gentamicin. Dilution was performed to assure single colonies. Plates were incubated at 30°C for 24 to 30 hours. Single colonies were simultaneously screened for purity by combinatorial colony hot-start PCR (see 2.5.3.2) and inoculated in LB with the respective antibiotic. This procedure was repeated until cultures were purified from cross-contaminations.

2.5.7 Genetic complementation

Mutants were for verified plasposon insertion (see 2.6.3.2) and stress sensitive phenotype as described by Reva *et al.* (2006) and candidate genes were complemented in *trans* ensure that the observed phenotype was caused by the particular plasposon insertion and thus inactivation of the gene, rather than by other secondary genetic events. If independent insertion events effected the same gene and respective mutants showed a consistent phenotype, correlation between plasposon insertion and observed phenotype were considered to be proven.

2.5.7.1 Generation of electrocompetent cells

Electrocompetent cells were generated following a protocol modified by Enderle & Farwell (1998). 2 mL of either *Escherichia coli* or *P. putida* overnight culture were sedimented by centrifugation at 6000 g for 10 minutes at 4°C. The bacteria pellet was washed twice with ddH$_2$O with subsequent centrifugation at 13000 g for 3 minutes at 4°C. After the last washing step, the pellet was resuspended in 200 μL pre-cooled ddH$_2$O and mixed with ~ 50 ng plasmid. The mixture was used for subsequent electroporation.

2.5.7.2 Electrotransformation

100 μL of the electrocompetent cell solution were transferred into pre-cooled electroporation cuvettes and cells were electroporated with the BioRad Pulser (440 W, 25 μF, 1.25 kV, time constant 5 ms) to induce plasmid uptake. The cell suspension was transferred to a fresh tube after adding 900 μL LB and incubated at 37 °C or 30°C, respectively with constant shaking (350 rpm) for 2 to 3 hours. Positive transformants or revertants were selected by spreading 100 μL onto LB plates containing the respective antibiotic and incubation at 37°C or 30°C overnight respectively. Single colonies were analysed by plasmid isolation and digestion.

2.5.8 Motility assays

The motility of bacteria in aquatic habitats is characterized by different forms of translocation. Within the genus *Pseudomonas* there are three types of motility: swimming, swarming and "twitching". Swimming and swarming motility are enabled by flagella, of which swimming is disorientated and swarming orientated. Twitching, the third type of motility is enabled by type IV pili (Mattick, 2002; Whitchurch *et al.*, 2005).

2.5.8.1 Swimming

For the verification of swimming motility, two μL of overnight culture were inoculated into 0.3% LB agar. Growth within the agar was examined after overnight incubation at 30°C by measuring the diameter of the colony.

2.5.8.2 Swarming

The capability of bacteria to swarm was examined by dropping two μL of overnight culture onto 0.5% LB agar and incubation overnight at 30°C. The diameter of the colony was measured to asses swarming motility.

2.5.8.3 Twitching

To examine the twitching motility of a bacterial strain, two µL of overnight culture were inoculated through the LB agar (1%) onto the surface of the petri dishes and incubated overnight at 30°C. The following day, the agar was removed and the plates were stained with Coomassie solution (0.05% Coomassie Brilliant Blue R250, 50% methanol, 10% acetic acid) in order to measure the twitching zone.

2.6 Molecular Biology Methods

2.6.1 Isolation of DNA

2.6.1.1 Isolation of genomic DNA from *Pseudomonas putida*

For genomic DNA isolation a modified method was used that was developed for gram-negative bacteria by Chen & Kuo (1993). 4 mL of overnight culture were harvested by centrifugation at 14000 g for 5 minutes. The bacteria pellet was resuspended in 900 μL lysis buffer and incubated for 20 to 30 minutes. Afterwards, 300 μL NaCl (5M) were added and the suspension was mixed vigorously. Proteins and cell fragments were separated by centrifugation (14000 g, 60 min, 4°C). The supernatant was transferred into a fresh tube. RNA was digested by addition of 15 μL DNase-free RNase A (10 mg/mL) and incubation at 37°C for 30 minutes. Purification of the DNA solution was achieved by consecutively mixing the aqueous phase with equal amounts of phenol, phenol/chloroform/isoamylalcohol (25:24:1) and chloroform/isoamylalcohol (1:24) with subsequent centrifugation (14000 g, 20-30 minutes, 4°C). If the aqueous phase was still turbid after phenol/chloroform extraction this step was repeated until the upper phase was clear. Precipitation of DNA resulted from adding an equal amount of isopropanol and slow rotation of the mixture. DNA was sedimented by centrifugation (14000 g, 3 minutes, RT). The pellet was washed with pre-cooled ethanol (-20°C) and dried at 37°C. Depending on the expected DNA amount the pellet was resuspended in 50 to 200 μL TE buffer or H_2O. To ensure that the DNA dissolves completely, the precipitate was stored for 1 to 2 days at 4°C.

2.6.1.2 Isolation of plasmid DNA

Plasmid DNA was isolated by a modified alkaline lysis method described by Birnboim & Doly (1979). Single colonies containing the constructed plasmid were picked and grown overnight at 30°C (*P. putida*) or 37°C (*E. coli*) in LB with the respective antibiotic. 5 mL of overnight culture were centrifuged (5000 g, 5 minutes) and the bacterial pellet resuspended in 300 μL solution I. Afterwards, 300 μL of solution II were added, mixed vigorously and incubated for 5 minutes at RT, following 300 μL of solution III and incubation on ice for 15 minutes. Cell debris were precipitated by centrifugation (10000 g, 10 minutes, 4°C) and the supernatant was transferred into a fresh tube. For protein removal a phenol/chloroform/isoamylalcohol (25:24:1) and chloroform/isoamylalcohol (24:1) extraction was carried out with subsequent centrifugation (14000 g, 20 minutes, 4°C). Plasmid DNA was precipitated with isopropanol as described above.

2.6.2 Quantification of nucleic acids

The concentration of nucleic acids in aqueous solutions was measured spectrophotometrically with ddH$_2$0 as standard by determining absorption at wavelengths 260 nm and 280 nm. The correlation is as follows:

$$\text{Absorption } A_{260nm} = 1 \rightarrow c_{(DNA)} = 50 \text{ ng/}\mu L$$
$$\rightarrow c_{(RNA)} = 40 \text{ ng/}\mu L$$
$$\rightarrow c_{(cDNA)} = 33 \text{ ng/}\mu L$$

The quotient A_{260nm} / A_{280nm} should be 1.8 – 2.0. Lower values indicate protein contamination.

2.6.3 Polymerase Chain Reaction

The polymerase chain reaction (PCR) is a method to amplify defined DNA fragments from a very small amount of DNA template.

The first idea to amplify DNA fragments with flanking oligonucleotides (primer) arose already in the early 1970s (Kleppe *et al.*, 1971), but not until the late 1980s when Kary Mullis (1986) reinvented the polymerase chain reaction, it became a standard method in molecular biology (Saiki *et al.*, 1988).

MATERIAL AND METHODS

There are three essential steps in this reaction:
1. Denaturation at 94°C to separate double-stranded DNA and eventually linked primer.
2. Annealing at 55 to 65°C to hybridize the primer to the complementary single-stranded DNA.
3. Elongation at 72°C: Hereby the desired DNA fragment is built due to elongation of the primer with nucleotides corresponding to the DNA template.

The construction of an efficient primer is a critical step for the PCR to obtain products with high yield and purity.

The following criteria were considered:
- GC-content of the primer should be comparable to the DNA template; in this case around 60-65% since *P. putida* has a high GC-content (61.5%).
- The melting temperature (T_M) should be above 60°C. This can be roughly estimated by the following formula (Marmur & Doty, 1962):

$$T_M = \sum (GC) * 4°C + \sum (AT) * 2°C$$

- Forward and reverse primer used in the same reaction should have approximately the same melting temperature.
- Optimal length of the primer is between 20 and 25 nucleotides.
- For optimal hybridization a primer should have neither palindromic structures nor four consecutive identical nucleotides.
- G or C nucleotides at the 3' end of the primer stabilize hybridization and facilitate elongation.
- Primer sequences should only show homology with the template DNA at the desired target sequence to avoid binding to other DNA regions and thus producing unspecific products.

Primer and corresponding sequences used in this study can be found in section 2.3.3.

2.6.3.1 Standard PCR

For standard applications *Taq*-polymerase (InViTek), in special cases such as sequencing and complementation GoldStar-*Taq* polymerase system (Eurogentec) was used. The GoldStar-*Taq* polymerase possesses higher proof-reading activity and is used when the incorporation of incorrect nucleotides has to be kept as low as possible (The error rate of InViTek-Taq polymerase is $2.7 * 10^{-5}$ according to the manufacturer). Since *P. putida* DNA has a high GC content (61.5 %) which supports formation of secondary structures. DMSO, a chaotropic reagent, was added to PCR reactions to facilitate efficient melting of the DNA strand.

The standard reaction volume was 50 µL and was prepared as shown below:

component	volume
10 x reaction buffer	5 µL
dNTPs (each 2 mM)	5 µL
DMSO	2.5 µL
MgCl$_2$ (50 mM)	1.5 µL
Primer (5 pmol/µL)	5 µL each
DNA template	50-100 ng
ddH$_2$O	Add to 50 µL
Taq-polymerase (5 U/µL)	0.2 µL

If other reaction volumes were required the volumina of the single compounds were accordingly adjusted. The PCR reaction mix was covered with a layer of paraffin since the lid of the thermocycler (Landgraf) cannot be heated.

For amplification the following programme was used accordingly to the primer properties and length of the DNA fragment. An approximate value for the elongation time (t_{elong}) is 60 sec per kb. The denaturation time (t_{den}) was normally set to 60 to 90 sec. In most cases the hybridization temperature (T_{ann}) of the primer was 60°C, in a few cases 58°C or 62°C.

Denaturation	300 sec		96°C	1 cycle
Annealing	60 sec		T_{ann}	
Elongation	t_{elong}		72°c	35 cycles
Denaturation	t_{den}		94°C	
Annealing	60 sec		T_{ann}	1 cycle
Elongation	2 x t_{elong}		72°C	
cooling	∞		10°C	

After the PCR reaction the product was verified by agarose gel electrophoresis (see 2.5.4) for size, quality and yield. If necessary an optimization of the PCR was achieved by independent variation of annealing temperature T_{ann} or the $MgCl_2$ concentration to affect hybridization of the primer to the DNA template or alter processivity (and error rate) of the polymerase.

In the case of very large products (> 2.5 kb) the FailSafe™ PCR system (Epicenter) was used. This system allows amplification of DNA strands up to 20 kb according to the manufacturer. Parallel reactions with different polymerases and 12 reaction buffers lead to a fast optimization of the PCR reaction.

2.6.3.2 Combinatory colony hot-start PCR

For the verification of purity and correct plasposon insertion mutants (see 2.3.1), a combinatory colony hot-start PCR was performed. Hereby, a combination of three primers were used, a primer pair flanking the plasposon insertion and one primer hybridizing to the gentamicin resistance cassette of the plasposon, resulting in three PCR reactions for each mutant and identifying at the same time the orientation of the inserted plasposon.

As a control the forward (F) and reverse (R) primer flanking the insertion site were used for PCR with template DNA from the *P. putida* wild type. PCR products gained from mutant or wild type DNA amplified with the flanking primer should differ by approximately 1.6 kb in size due to the inserted plasposon.

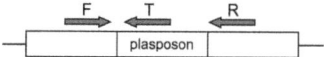

In case of colony PCRs the DNA template was provided by entire bacterial cells instead of purified DNA solutions.

An aliquot of 200 µl of overnight bacterial culture was centrifuged for 3 min at 13000 *g* at RT. The pellet was resuspended in 1 ml ddH_2O. One µl thereof was subjected to combinatorial colony hot-start PCR in a total volume of 25 µl. Polymerase was added after the initial denaturation step, wherefore a breakpoint at 60°C was integrated in the PCR program. PCR products were separated by agarose gel electrophoresis afterwards.

component	volume
10 x reaction buffer	2.0 µL
dNTPs (each 2 mM)	2.5 µL
DMSO	1.25 µL
MgCl$_2$ (50 mM)	0.75 µL
Primer (5 pmol/µL)	2.5 µL each
Cell suspension	1 µL
ddH$_2$O	7.5 µL
10 x reaction buffer	0.5 µL
ddH$_2$O	4.3 µL
Taq polymerase (5 U/µL)	0.2 µL

Denaturation	300 sec	96°C	1 cycle
Cooling to 60°C annealing temperature			
Breakpoint			
Annealing	60 sec	T$_{ann}$	35 cycles
Elongation	t$_{elong}$	72°C	
Denaturation	t$_{den}$	94°C	
Annealing	60 sec	T$_{ann}$	1 cycle
Elongation	2 x t$_{elong}$	72°C	
cooling	∞	10°C	

2.6.4 Agarose gel electrophoresis

Agarose gel electrophoresis can be used for separation of nucleic acids for analytical and preparative purposes. In this study agarose gel electrophoresis was used to determine size, quality and quantity of PCR products, genomic DNA or total RNA.

Depending on the expected size of the DNA fragments gels with concentrations of 1 to 3% agarose in TBE buffer were prepared (sizes: 5 x 7 cm, 10 x 14 cm or 20 x 20 cm). TBE buffer was used as electrophoresis buffer. Before samples were applied to the gel, aliquots were mixed with loading buffer (1%). Electrophoresis was applied at room temperature with field strength of 5-8 V/cm (Powerpac 300 Power Supply, BioRad), for slow separation at 4°C with 1.5-2 V/cm.

After appropriate separation time the gel was stained in 0.5 µg/mL ethidium bromide solution for 30 minutes.

Ethidium bromide incorporates in between the bases of the DNA and thereby stains the DNA. Afterwards excessive ethidium bromide was removed in two washing steps in H$_2$0, each for 20 minutes. DNA was visualized with UV-light (wavelength 312 nm) on a UV-transilluminator (Bachofer) and photographed for documentation.

A standard DNA ladder was applied together with the DNA samples to estimate the size of the separated DNA fragments. Three different markers were used with the following fragment sizes (bp):

λDNA-BstEII digest (NEB):
8454, 7242, 6369, 5686, 4822, 4324, 3675, 2323, 1929, 1371, 1264, 702, 224, 117

100 bp DNA ladder (Fermentas):
1000, 900, 800, 700, 600, 500, 400, 300, 200, 100

50 bp DNA ladder (Fermentas):
1000, 900, 800, 700, 600, 500, 400, 300, 250, 200, 150, 100, 50

2.6.5 Isolation of DNA fragments from agarose gels

This procedure was used to extract DNA fragments of interest after electrophoresis from the agarose gel. The PCR product was visualized and the respective gel excised with a scalpel. The purification of the product was carried out with the QiaQuick GelExtraction Kit (Qiagen) following the manufacturer's instructions.

2.6.6 DNA preparation for genetic complementation

For complementation *in trans* candidate genes together with their promoter region were amplified from *P. putida* KT2440 DNA by PCR using the GoldStar-*Taq*-Polymerase (see 2.5.3.1). PCR products were separated by agarose gel electrophoresis (see 2.5.4), purified and ligated into the pUCP20 vector. Ligation products were then introduced into target cells by electroporation (see 2.5.7.2).

2.6.6.1 Restriction digestion of DNA

The restriction digestion mixture was as follows:

Digestion mix	Insert DNA	pUCP20
DNA	x µL (5 µg DNA)	x µL (5-10 µg DNA)
Buffer NEB 2	3 µL	3µL
10x BSA	3 µL	3 µL
Enzyme 1 SphI/XbaI	1 µL/1.5 µL	1 µL/1.5 µL
Enzyme 2 HindIII /HindIII	2 µL/1.5 µL	2 µL/1.5 µL
ddH₂O	add to 30µL	add to 30 L

The mixture was incubated overnight at 37°C and the digested DNA was again purified using the QiaQuick GelExtraction Kit.

The digested vector DNA was dephosphorylated in two steps to remove 5'end phosphate groups and thus prevent self-ligation. This step can be used to remove vector background in cloning strategies (Sambrook *et al.*, 1989).

Dephosphorylation mix	Step I	Step II
Purified vector DNA	80 µL	100 µL
Buffer NEB3	10 µL	1 µL
CIP phosphatase	5 µL	5 µL
ddH2O	5 µL	4 µL
Final volume	100 µL	110 µL
Incubation	37°C for 60 min	37°C for 30 min
Inactivation	80°C for 10 min	80°C for 5 min

DNA samples were ready for ligation and stored at -20°C prior to use.

2.6.6.2 Ligation

Vector and insert DNA were mixed in proportion 1:3 to assure maximal ligation efficacy. Two µL T4 buffer, 2 µL ATP (10 mM), 1 µL T4 ligase and ddH2O were added to give a final volume 20 µL. The ligation mixture was incubated for 2 hours at 25 °C and subsequently used for electroporation.

2.6.7 DNA:DNA Hybridization

This method is used for the identification and further analysis of DNA fragments. Hereby, nucleic acids that are separated by gel electrophoresis are transferred to a membrane and immobilized. The easiest transfer method is capillary transfer which was used in this study. After the adjacent hybridization step the DNA fragments can be detected and analyzed. Generally, Southern-Blotting is known as the transfer and fixation of nucleic acids to a membrane (Southern, 1975).

2.6.7.1 Generation of digoxigenin-labelled DNA probes

Probe sequences were amplified by PCR (see 2.5.3.1), and isolated and purified from agarose gels (see 2.5.4 and 2.5.5). In the following, labelled DNA fragments were generated by incorporating digoxigenin-linked nucleotides. Therefore, 10 µL of DNA solution and 5 µL ddH$_2$O were heated at 94°C for 4 minutes and immediately cooled on ice. After addition of 2 µL hexanucleotide-mix, 2 µL 10x DIG-dUTP-Labelling mix and 1 µL Klenow polymerase, the reaction mix was incubated at

37°C overnight and afterwards mixed with 280 µL TE buffer and a dye mix (0.8% dextran blue (2×10^6 g/mol) and 0.5% phenol red (376 g/mol) dissolved in TE buffer).
Before usage the DNA probe was purified by gel filtration through Sephadex G-50 columns (Ø 5 mm, h 4 cm, Pharmacia Fine Chemicals). The column was washed with 2 mL TE buffer and dried by centrifugation (1000 g, 45 sec). The DNA probe was then pipetted onto the gel column and collected by another centrifugation step (1000 g, 30 sec). All unused hexanucleotides, dNTPs and phenol red were retained on the Sephadex column. The DNA probe was stored at -20°C prior to use.

2.6.7.2 Restriction digestion of genomic DNA

For DNA:DNA hybridization genomic DNA was digested with a restriction enzyme that cuts the genomic DNA of *P. putida* relatively often, but not in the DNA fragment of interest. In this study, the enzymes BamHI and PstI (NEB) were used. The reaction mix was incubated at 37°C overnight.

Digestion mix	
Genomic DNA	5 µg
10 x reaction buffer	2 µL
100 x BSA (if required)	0.2 µL
Enzyme	2 µL
ddH$_2$O	add to 20 µL

The restriction digestion was examined by agarose gel electrophoresis. The total reaction volume was loaded onto a 1% agarose gel (20 x 20 cm) and separated (1.5 V/cm) at 4°C for 24 hrs. The gel was stained with ethidium bromide, destained and photographed for documentation. Afterwards the DNA was transferred to a nylon membrane (see next part).

2.6.7.3 Transfer and fixation of DNA to a membrane (Southern Blot)

The Southern Blot was performed using a capillary transfer under alkaline conditions (0.4 M NaOH) to transfer and locate the DNA from an agarose gel to a nylon membrane (Hybond N$^+$ membrane, Amersham Pharmacia). The gel was placed onto Whatman paper that had contact with the transfer buffer (0.4 M NaOH) reservoir with both ends to ensure the capillary effect. The nylon membrane was set on the gel and covered by two layers of Whatman paper. Meanwhile, every single layer was soaked with transfer buffer to avoid air bubbles. For an efficient capillary effect, a thick layer of absorbent paper was placed on top of the construction. The transfer lasted 12 to 16 hours. Afterwards, the membrane was neutralized by washing twice for 5 minutes with neutralization buffer and air-dried. DNA was immobilized onto the membrane by UV irradiation (UV-Crosslinker, Stratagene). If required the membrane can be stored at -20°C prior to use.

2.6.7.4 Hybridization and detection of digoxygenin-labelled DNA

The hybridization reaction allows the detection of specific membrane-bound DNA fragments with digoxigenin labelled DNA probes.

The membrane was placed in a hybridization tube that was filled with prehybridization buffer and incubated in a rotary oven (Bachofer) at 68°C for at least 4 hrs. Afterwards, the labelled DNA probe was dissolved in 10 mL prehybridization buffer and poured onto the membrane which was incubated at 68°C overnight. The probe could be recovered and stored at -20°C. The membrane was washed at 68°C at least 2 x 20 min in washing buffer to remove unbound probe.

The detection of the DIG-DNA labelled fragments was conducted according to the protocol by Guiliano et al. (1999). The membrane was placed in a plastic basin and treated as follows:

1.	Equlibration	Buffer I	150 mL	5 min
2.	Blocking	Buffer II	200 mL	30 min
3.	Incubation	Antibody solution	50 mL	30 min
4.	Washing	Buffer I	200 mL	3 x 20 min

In the third step, an anti-DIG antibody that is linked to the alkaline phosphatase, binds to the DIG-labelled probe on the membrane.

The membrane was then transferred to a glass basin that was beforehand washed with buffer III.

5.	Equlibration	Buffer III	20 mL	5 min
6.	Incubation	CDP Star solution	10 mL	3 min

In the last step CDP Star is cleaved by the alkaline phosphatase and thus chemoluminescence is released that can be detected by X-ray exposure (X-ray film, X-OMAT^{TM} AR or Bio-MAX MR, Kodak). Before X-ray exposure, the membrane was laminated.

2.6.7.5 Regeneration of hybridized DNA membranes

DIG-DNA labelled membranes can be recovered for further experiments. For this, they are treated by two washing steps (each for 30 min in Stripping buffer).

The linkage between digoxigenin and dUTP is unstable in alkaline solutions. Thus, the bondage breaks and digoxigenin and the bound antibody are released.

Afterwards, the membrane was washed intensively in H_2O to remove SDS, neutralized in blot washing buffer for 5 minutes, laminated and stored at -20°C.

This procedure enables the removal of the DIG-DNA labelling, but not of the DNA fragments from the probe. For this reason, the membrane cannot be regenerated endlessly, as the number of available binding sites for new fragments decreases with every usage.

2.6.8 The Y-linker method

The Y-linker method can be used to identify flanking genes of a plasposon insertion as it is independent of the insertion sequence composition. It is PCR-based and depends on the ligation of a designed 40 bp oligonucleotide (Y-linker) to randomly digested mutant chromosomal DNA (200 – 1200 bp). By performing a subsequent PCR with a primer (Y-primer) specific to the non-complementary part of the Y-linker combined with a plasposon specific primer, a PCR product is generated and can be sent for sequencing to identify DNA regions the flanking the insertion (Kwon & Ricke, 2000).

2.6.8.1 Generation of the Y-linker

The Y-linker was generated by ligation of two oligonucleotides (linker strand 1 and 2, see 2.3.3), whereof one part was complementary and built a double-strand and the second part non-complementary and built the y-shape. Linker strand 2 was dephosphorylated by incubating the reaction with a T4-PN-kinase mix at 37°C for 15 minutes and subsequent heating to 95°C for 20 minutes (Thermomixer comfort, Eppendorf). Then, strand 1 was added and the volume was adjusted to 80 µL. The heater was switched off and the ligation mixture was incubated until room temperature was reached slowly. The two y-linker strands annealed and formed the Y-linker. Finally, another 40 µL ddH$_2$0 were added.

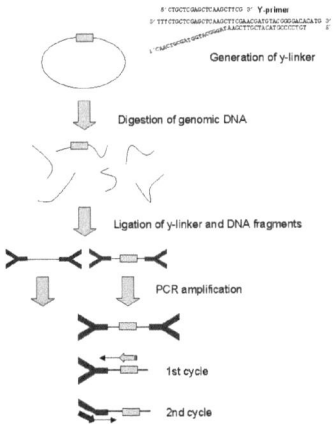

Fig. 2.2 Schematic overview about the Y-linker method to generate a PCR product and to identify the flanking region of a plasposon insertion by sequencing (Weinel, 2003; modified).

Reaction mix	volume	
strand 2 (3.5 µg/µL)	4 µL	
ATP (10 mM)	4 µL	
10 x T4 PNK buffer (NEB)	4 µL	Phosphorylation
4 PN kinase	1 µL	
dd H$_2$O	27 µL	
37°C for 15 min and 95°C for 20 min		
Strand 2 (3.5 µg/µL)	4 µL	Annealing
ddH20	36 µL	
Cooling to RT		
ddH2O	40 µL	

2.6.8.2 Y-linker Ligation

Genomic DNA from mutant strains (1 µg) prepared as described in chapter 2.5.5.1 was digested with either NlaIII (5 U) or Sau3AI (5 U) in a final volume of 20 µL. The reaction was incubated at 37°C for 3 hours and immediately used for ligation. For the ligation about 500 µg digested genomic DNA and 5 µL of the annealed Y-linker were incubated at RT for 2 hours together with 2 µL ATP (10 mM), 2 µL T4 buffer and 1 µL T4 ligase (Fermentas) in a final volume of 20 µl adjusted with ddH$_2$O. Approximately 50 ng of the ligation mixture were used as template for PCR using the Y-primer and the specific plasposon primer afterwards. The PCR product was purified and sent for sequencing.

2.6.9 DNA sequencing

For the identification of the flanking gene of the plasposon insertion, the Y-linker PCR product was purified via agarose gel electrophoresis as (described in chapter 2.6.4 and 2.6.5.) and then sent for sequencing. Sequencing was performed by Qiagen (Qiagen Sequencing Services, Hilden). For sequencing the same primers were used as for generation of the PCR product.

2.7 RNA Working Technique and Transcriptome Analysis

2.7.1 RNA handling

All solutions, consumables and equipment were autoclaved at 121°C for 60 minutes to avoid contamination with bacterial RNA. Equipment that was not autoclavable was sterilized with 70% ethanol and cleaned with ddH$_2$O, glassware heat-treated at 250°C for 5 hours. All aqueous solutions were prepared with 0.05% DEPC to inactivate RNases and subsequently autoclaved at 121°C for 30 minutes to ensure the complete decomposing of the DEPC. Tris- and acetate-containing solutions were prepared with DEPC-treated H$_2$O. Chambers for agarose gel electrophoresis were incubated with 3% H$_2$O$_2$ for 15 minutes and afterwards washed with 70% ethanol.

2.7.2 RNA extraction and storage

Total RNA was extracted according to the protocol provided by Qiagen (RNeasy Mini Kit).
Bacterial cultures, three in parallel, were grown until the desired cell density was reached. For cell harvest, 2 volumes RNAprotect Bacteria Reagent (Qiagen) were added to one volume bacterial culture and mixed vigorously. The solution was incubated at RT for 5 min and immediately centrifuged at 5000 g for 10 minutes. The supernatant was decanted and if required the bacterial pellet could be stored at -20°C for up to 2 weeks or at -80°C for up to 4 weeks.
For cell lysis, always the required amount of TE buffer containing 1 mg/mL lysozyme was added; the cell pellet was resuspended and incubated at RT for 20 min. In this study, 1 mL TE buffer was added to 10 mL bacterial culture with an OD$_{600nm}$ ~ 0.8. Afterwards, 1800 µL RLT buffer containing 1% β-mercaptoethanol were added, mixed intensively, followed by 1200 µL EtOH.
The RNA solution was then purified using the RNeasy Mini Kit, thereby applying the total volume to one column. The optional on-column DNase digestion as described in the protocol was performed twice for 20 min to ensure the complete removal of genomic DNA. Afterwards, RNA was eluted with 30 µL RNase-free H$_2$O, treated with 2 µL RNase inhibitor (SUPERaseIn, Ambion) and could be stored at -80°C for up to 4 weeks.

2.7.3 Formaldehyde agarose gel electrophoresis

RNA purity, quality and quantity were controlled by formaldehyde agarose gel electrophoresis. Formaldehyde allows the electrophoretic separation as it keeps RNA in a denatured condition. Therefore, agarose gel and RNA loading buffer were prepared with formaldehyde.

The agarose gel was prepared in MOPS buffer, also used as electrophoresis buffer, and had an agarose concentration of 1.2%. Before the gel was poured formaldehyde was added to an end concentration of about 2.1%. Two µL of RNA sample were denatured at 65°C for 10 min, cooled on ice, mixed with 5 µL formamide, 2 µL formaldehyde, 1 µL 10x MOPS and 2 µL RNA loading buffer and transferred to the gel. Electrophoresis was performed with 5 V/cm at RT under a fume hood, since formaldehyde is volatile. The gel was stained with the procedure described in chapter 2.5.4.

Only 16S rRNA and 23S rRNA were detectable as distinct bands, on the basis of which quality and purity of the RNA was estimated.

2.7.4 cDNA generation

cDNA generation was performed with a statistically distributed mixture of hexanucleotides as primers (random priming), whereby the total RNA in one sample is transcribed into cDNA. In this study cDNA was generated using SuperscriptII reverse transcriptase (Invitrogen) according to the manufacturer's protocols.

For the cDNA synthesis about 10 µg total RNA, pooled from three independent RNA samples, were applied. The RNA pool was adjusted to 14.5 µL either by adding RNase-free H_2O or concentration with MicroCons YM-10 columns (Millipore) at 6000 g at 4°C for up to 15 min.

RNA pool (10 µg]	14.5 µL
Random hexamer p(dN_6), 10 µg/µL, Roche)	1.5 µL

The RNA hexamer mixture was incubated at 70°C for 10 min, immediately cooled on ice and centrifuged (6000 g, 10 sec). Then a dNTP mixture was added.

RNA mix	16 µL
5x 1st strand buffer	6 µL
DDT (0.1 M)	3 µL
RNasin	1 µL
dNTPs (40 mM)	1 µL

After incubation at 25°C for 2 min, 3 µL reverse transcriptase (Superscript II, 200 U/µL, Invitrogen) were added and cDNA was synthesized at 42°C for 2 hrs.

The reaction was stopped by adding 10 µL EDTA (0.5 M) and residual RNA was removed by 10 µL NaOH (1M) at an incubation temperature of 65°C for 10 min. Afterwards, 25 µL HEPES (1 M; pH 7.5) and at least 25 µL sodium acetate (3 M, pH 5.2) were added. The pH of the cDNA samples was checked prior purification, since DNA can only bind to the columns when pH is slightly acidic. The cDNA samples were purified using the QiaQuick PCR Purification Kit (Qiagen) following the manufacturer's instruction.

The protocol provided by Progenika was optimized for indirect labelling, thus they used aa-dUTPs. cDNA synthesis was performed as described above with one minor modification. Instead of a 40 mM dNTP mix, a 30x dNTP/aa-dUTP mix was used.

For the preparation of dNTP/aa-dUTP aliquots 1 mg aa-dUTP was dissolved in 17 µL ddH$_2$O and 0.68 µL NaOH (1M). To prepare 30x dNTPs, dNTPs with the following concentration were mixed and adjusted to 140 µL total volume:

	stock concentration	volume	end concentration
dATP	78.96 mM	26.6 µL	15 mM
dCTP	82.07 mM	25.58 µL	15 mM
dGTP	77.02 mM	27.27 µL	15 mM
dTTP	80.1 mM	5.24 µL	3 mM
aa-dUTP	100 mM	17.0 µL	12 mM
ddH2O		38.51 µL	

2.7.5 RT/PCR

For RT/PCR of selected genes, firstly 2 µg of total RNA, derived from three independent samples, were used for cDNA generation with 2 pmol of gene specific reverse primers specific for first strand generation. As for general cDNA synthesis, cDNA was generated using SuperscriptII reverse transcriptase (Invitrogen) according to the manufacturer's protocols.

cDNA synthesis mix	
total RNA	2 µg
reverse primer	2 pmol
H$_2$O	add to 10 µL
5 minutes at 70°C	
5x 1st strand buffer	4 µL
DDT (0.1 M)	2 µL
RNasin	1 µL
dNTPs (40 mM)	2 µL
2 minutes at 42°C	
Superscript II	1 µL
60 minutes at 42°C	
10 minutes at 70°C for heat inactivation of enzymes	

For subsequent PCR reactions, 2 µL of the cDNA reaction mix were used as template. PCR reactions were performed as described in 2.6.3.1. To assess transcript amount, aliquots for agarose gel electrophoresis were taken first after 10 cycles, and then following every third PCR cycle up to cycle 25. The amount of transcript (amol RNA) was calculated due to appearance of first detectable PCR product according to the protocol by Bremer *et al.* (1992).

2.7.6 cDNA labelling

2.7.6.1 cDNA labelling for *Pseudomonas putida* Genome Arrays (Progenika)

The *P. putida* genome oligonucleotide array (Progenika Biopharma, S.A.) is a two-dye array, using the fluorescent dyes cyanine-3 (Cy3) and cyanine-5 (Cy5) (GE Healthcare). Both samples, reference and test samples, are hybridized to one array. For this reason reference and test sample had to be prepared simultaneously. In this study, 4 arrays per experiment were used in parallel to ensure statistically reliable results. Therefore, each sample was labelled twice with each dye, resulting in eight cDNA samples per experiment.

Before labelling, Cy3 and Cy5 aliquots were prepared by dissolving the substance in 40 µL DMSO, making aliquots of 4 µL each and lyophilizing the label in a vacuum concentrator. Furthermore, aliquots of 4 M hydroxylamine were prepared and stored at -80°C prior to use.

During all steps of cDNA labelling and adjacent hybridization, fluorescent dyes and labelled cDNA should be protected from light as the dye is light sensitive.

At least 5 µg of total cDNA per sample were used for the labelling reaction and at first concentrated in a vacuum concentrator at 45°C until all liquid evaporated. The pellet was then dissolved in 9 µL

NaHCO$_3$ (0.1 M, pH 9.0). Cy3 or Cy5 aliquots were added, mixed and incubated at RT for 2 h in the dark with shaking and spinning down every 20 min. If necessary, samples could be stored at -20°C overnight before quenching. The dye was quenched with 4.5 μL hydroxylamine (4 M) and incubated again in the dark for 15 min. The samples were adjusted to a total volume of 75 μL with ddH$_2$O and purified using the PCR purification columns provided by Qiagen, but with self-made buffers for the washing and elution step. cDNA was bound to the columns with the commercial binding buffer (Qiagen) and then washed twice with 750 μL washing buffer (potassium phosphate 4 mM, 80% ethanol, pH 8.0) and eluted with 435 μL elution buffer (potassium phosphate, 5 mM, pH 8.5). An aliquot was used for quantity and quality control of cDNA (see 2.5.2) and measuring labelling efficiency.

Cy3 absorbs at 550 nm, Cy5 at 650 nm. Minimum sample requirement is 130 pmol of dye incorporation. Labelling efficiency can be calculated as follows:

$c = A/(\Sigma * l)$ c (label-efficiency) in pmol/μl

A: absorbance
Σ: extinction coefficient
$\Sigma_{Cye3} = 150.000$ M^{-1} cm^{-1}
$\Sigma_{Cye5} = 250.000$ M^{-1} cm^{-1}
l: dilution / layer thickness

Equal amounts of cDNA of the samples to be compared on one array were mixed and dried in a vacuum concentrator at 40°C for several hours.

2.7.6.2 cDNA fragmentation and labelling for Affymetrix microarrays

For hybridization on Affymetrix microarrays, cDNA samples were fragmented and terminally labelled according to the protocol provided by Affymetrix.

For the fragmentation, 3 μg total cDNA were incubated with 2 μL 10x One-Phor-All Buffer and DNaseI (0.6 U/μg) in a total volume of 20 μL (adjusted with ddH$_2$O) and incubated at 37°C for 10 min. The reaction was stopped by inactivating the enzyme (98°C, 10 min) and samples were cooled on ice. Then, the resulting fragment sizes were controlled by gel electrophoresis (5-8 V/cm, 1 h) on a 3% agarose gel and thereafter stained with SYBR Gold. About 300 ng cDNA were mixed with 8 μL sucrose dye and applied to the gel. A 50 bp DNA ladder was used as molecular size standard. For optimal labelling and hybridization results, cDNA fragments should have a size between 50 and 200 nucleotides.

The total amount of fragmented cDNA was then used for the terminal labelling reaction. The labelling mix was incubated at 37°C for 1 h and the reaction was stopped by adding 2 μL EDTA (0.5 M).

Labelling mix	
Fragmented cDNA	
5x reaction buffer	10 μL
Gene Chip DNA Labeling	2 μL
Terminal tranferase	2 μL
ddH$_2$O	add 50 μL

The efficiency of the labelling reaction was assessed by a gel-shift assay to prevent hybridization of poorly labelled samples.

The method is based on the very strong binding affinity of avidin (also neutravidin) to biotin. The DNA-protein complex is much larger than the fragmented cDNA and thus less mobile during electrophoresis. A shift pattern assures good biotin labelling.

For each sample to be tested, 2 aliquots (each ~200 ng) were removed, one of which was mixed with 5 μL NeutrAvidin solution (2 mg/mL) and incubated at RT for 5 min. Gel electrophoresis was run as described above.

2.7.7 Microarray blocking and hybridization

Labelled cDNA was sent to the Array Facility at the HZI, Braunschweig, for array hybridization and data evaluation processes.

2.7.7.1 Blocking and hybridization for *Pseudomonas putida* genome arrays (Progenika)

Prior to hybridization it is necessary to block the surface of the array to reduce background noise by non-specific binding of nucleic acids. Array slides were incubated in preheated blocking buffer at 42°C and after 45 min washed several times by dipping into ddH$_2$O followed by a final washing step in isopropanol. Slides were dried by centrifugation (2000 g, 5 min) and subsequently used for hybridization.

The desiccated cDNA samples were resuspended in 110 μL hybridization buffer and denatured at 95°C for 2 min. Meanwhile, the array slide was placed in the hybridization chamber and preheated to 42°C. A coverslip was placed on top of the slide with white spacers facing down and the cDNA-solution was added slowly in between slide and cover until the array area beneath was completely covered. Samples were hybridized at 42°C for 16 hrs.

Hybridized arrays were treated by several washing steps with subsequent centrifugation (2000 g, 5 min). The first washing step was done with preheated (42°C) washing buffer 1, all other steps at RT:

- washing buffer 1   2x 5 min
- washing buffer 2   2x 5 min
- washing buffer 3   2x 5 min
- washing buffer 4   15 sec

2.7.7.2 Hybridization of Affymetrix microarrays

The cDNA samples prepared for Affymetrix microarrays were hybridized by mixing 55 µL cDNA sample and 155 µL hybridization mix with subsequent incubation at 50°C for 16 h at 60 rpm (Affymetrix GeneChip Hybridization Oven 640):

Hybridization mix	
2x hybridization buffer	100 µL
Control Oligo B2(900301, Affymetrix)	3.4 µL
Salmon Testis DNA (D7656, SIGMA)	2 µL
BSA (15561-020, Invitrogen)	2 µL
DMSO	14.2 µL
Nuclease-free H_2O	23.4 µL

Afterwards, the hybridization solution was replaced by 250 µL buffer A and then B and arrays were transferred to the Affymetrix GeneChip Fluidic Station 450 (program used: Midi_euk-2v3) for subsequent washing and staining sequentially with 600 µL SAPE-, Antibody- and again SAPE-solution. Then, microarrays were scanned using the Affymetrix GeneChip Operation Software Version 1.4 (Affymetrix GeneChip Scanner GCS 30) and data provided for further analysis.

2.7.8 Microarray data elevation

Data were taken up by the Affymetrix GeneChip Scanner GCS 30 run with Agilent Scan Control Software. For evaluation of signal intensities the Agilent Feature Extraction Software was used that calculated the signal intensity per spot. Scan pictures had a resolution of 5 µ per pixel. For Progenika and Affymetrix arrays different gene expression features were chosen according to the respective microarray design, for Progenika arrays the two channel gene expression feature was used, for Affymetrix microarrays the one channel feature. Signal intensity per spot was calculated by subtracting background from foreground and provided as median value of all measured pixel per spot.

Progenika array data were normalized using the LOWESS (Locally weighted linear regression) to remove intensity-dependent effects that can appear by low signal intensities and to avoid a bias due to different label efficiencies of the fluorescent dyes (Quackenbush, 2002).

Affymetrix arrays were median scale normalized providing a normalization factor for each array. This factor should be similar between arrays both within an experiment as a control for technical bias or between different experiments to reliably analyse gene expression profiles under different conditions.

Raw data were provided by the Array Facility and analysed with the freely available Software BRB ArrayTools (http://linus.nci.nih.gov/pilot/index.htm). Signal intensities were $\log_2$ transformed and median normalized over all arrays of one experiment, and an output file was generated providing mean signal intensities, p-value (two-sample T-test (with random variance model)), FDR (false discovery rate) and fold-change for all represented probe sets. Genes were considered to be significantly differentially expressed if in combination p-value ≤ 0.05 and FDR ≤ 0.05.

2.8 Illumina cDNA Sequencing

The Illumina sequencing technology allows parallel sequencing of millions of fragments based on reversible terminator chemistry. Randomly fragmented DNA is attached to a planar, optically transparent surface on which the fragments are extended and bridge amplified. In this way a high density flow cell is created with millions of clusters that contain several hundred copies of the same template. These templates are then sequenced by a four-colour DNA sequencing-by-synthesis technology applying reversible terminators with removable fluorescent dyes.

In addition to the transcriptome analysis using Progenika and Affymetrix arrays, *P. putida* wt cDNA from the very same samples was analyzed using the Illumina genome analyzer sequencing technology. For this, cDNA was generated and purified as described in chapter 2.7.4.2 and two samples (each 25 µg), one for each condition, were sent for sequencing to GATC Biotech AG (Konstanz).

In summary, the cDNA was nebulized to generate fragments less than 800 bp long. A terminal 'A' was then transferred to the 3'end and cDNA fragments were ligated to adapters, purified and bridge amplified. 36 cycles of sequencing–by-synthesis were performed for each library with the Genome Analyzer GAII SR generating millions of 36 bases reads. The Illumina Genome Analyzer Pipeline V2.0 software with default signal quality filters was used to provide each nucleotide of a read with a quality score Q_{solexa} and cohesively a probability value $p(X)$.

The correlation of quality score and probability value is: $Q_{solexa} = 10 \log (p(X) / 1-p(X))$. Thus, a maximal quality score (Q_{solexa}) of 40 indicates a probability of 99.9% for a correct base prediction/identification.

Lower quality scores typically appear at the end of reads, thus generally four bases were trimmed from the 3'end before further analysis.

2.9 Metabolome Analysis

The analysis of the intracellular metabolome profile of *P. putida* wt and mutants was accomplished in collaboration with Christian Jäger from the Department of Bioinformatics and Biochemistry at the Technical University of Braunschweig. First steps of sample preparation including extraction were performed in Hannover. Samples were desiccated and sent to Braunschweig for final processing, measurement and initial analysis.

Bacterial cultures as for transcriptome and proteome were grown until desired cell density was reached and cells were harvested according to 16 mg cell dry weight. For determination of cell dry weight see chapter 2.5.4. For each experiment 6 samples were processed in parallel (technical replicates). During the entire preparation, samples were kept on ice.

Cells were harvested and sedimented. Bacterial pellets were washed twice in 20 mL 0.9 % NaCl with subsequent centrifugation. All centrifugation steps were done at 4630 g for 3 min at 4°C. For cell extraction, supernatant was discarded completely after the last washing step and the pellet was resuspended in 1.5 mL pre-cooled (4°C) methanol containing 60 µL ribitol (0.2 g/L) as standard for metabolite concentration evaluation. The extraction mixture was incubated in an ultrasonic bath at 70°C for 15 min for cell lysis and afterwards cooled on ice for 2 min. Then cells were extracted by adding 1.5 mL ddH2O and 1 mL chloroform consecutively with vigorous shaking in between, followed by a final centrifugation step at 9000 g for 5 min at 4°C. The upper aqueous phase was transferred in a fresh tube and desiccated with inert nitrogen. Samples were stored at -80°C until being sent to Braunschweig.

Before GC-MS analysis, samples were derivatized by adding 20 µL pyridine containing methoxyamine hydrochloride (20 mg/mL), mixed for 20 sec and incubated at 30°C for 90 min with constant shaking at 600 rpm. Then, 32 µL N-methyl-N-trimethylsilyltrifluoroacetamide (MSTFA) were added and the reaction mixture was incubated for 30 min at 37°C, followed by 120 min at 25°C and finally centrifuged at 18400 g for 3 minutes. 50 µL of the derivate were transferred in a glass vial with micro cartridge for GC-MS analysis (Finnigan Trace GC + MS, Thermo Scientific) with the following specifications:

MATERIAL AND METHODS

Finnigan Trace GC + MS settings	
Autosampler	AS 3000 (Thermo Scientific)
Injector	PTV (*programmed temperature vaporizer*)
Initial temperature / evaporation time	70 °C / 0.2 min
Transfer ramp	14 °C sec^{-1}
Final temperature / final time	280 °C / 5 min
Clean rate / clean temperature / clean time	14 °C s-1 / 330 °C / 10 min
Injector mode	split
Split ratio	25 : 1
Sample volume	1 µl
Syringe	Thermo syringe
Volume (gauge/length/point style)	5 µl (26/50/Cone)
Column (*fused silica*)	Agilent® J&W Scientific DB-5MS
Length, ID, film thickness	30 m, 0.25 mm, 0.25 µm
Temperature range (isothermal limit/temperature program limit)	-60 ... 330/350 °C
Stationary phase	low polarity phase 5% diphenyl / 95% dimethylpolysiloxane
Column flow regulation	DPFC (Digital Pressure and Flow Control)
Gas flow column / Linear velocity (temperature)	1 ml min^{-1} ≙ 37 cm sec^{-1} (@ 70 °C)
Gas flow total (septum purge flow - fixed)	31 ml min^{-1} (5 ml min^{-1})
Gas saver	ON (10 ml min^{-1} after 6 min)
Vacuum compensation	ON
Carrier gas	Helium 5.0
Oven settings	temperature program
Initial temperature	70 °C
Initial time	1 min
Ramp 1	1 °C min^{-1}
Final temperature 1	76 °C
Final time 1	0 min
Ramp 2	6 °C min^{-1}
Final temperature 2	325 °C
Final time 2	10 min
Acquisition time	58.5 min
Transfer line temperature	275 °C
Ion source	*E*lectron *I*onization (EI)
Ionization energy	70 eV
Ion source temperature	200 °C
Analyser / detector	Quadrupole / Photomultiplier
Type of acquisition	Full scan
Acquisition rate	2.5 scans sec^{-1}
Mass spectrum	40 ... 460 amu
Start acquiring after eluting solvent	4.5 min
Measurement software	Xcalibur 1.2 (Thermo)

For data analysis the freely available software MetaboliteDetector (http://md.tu-bs.de/) was used.

2.10 Proteomic Analysis by LC-ESI-MS/MS

Bacterial cultures were grown to the desired OD as for transcriptome and metabolome analysis. Centrifugation tubes containing 0.5x volume killing buffer of required sample volume were pre-cooled at -20°C. Cells (40 OD_{500} units) were harvested quickly at 5000 g for 3 min at 4°C, supernatant discarded and cell pellet frozen in liquid nitrogen and stored at -80°C.

Further sample preparation and mass spectrometric analysis were carried out in the Interfaculty Institute for Genetics and Functional Genomics, Department of Functional Genomics, Ernst-Moritz-Arndt-University of Greifswald, in accordance to the following protocol kindly provided by Frank Schmidt.

Sample preparation: 10 µg protein of each sample (in 1x UT-solution) was adjusted to a final volume of 1.3 µL using 1x UT-solution to have similar concentrations and volumes. Samples were diluted 1:10 with 50mM bicarbonate-solution to reduce UT-concentration and to assure a basic pH for good trypsin performance. 20 µl of trypsin-solution (10 ng/µl 20 mM bicarbonate) was added and the samples were incubated at 37 °C for 15 h. To stop digestion 6.6 µl of 5 % acetic acid (ultra pure) was added. ZipTips (C18, Millipore) were sequentially washed 3× with 10 µl acetonitrile (ACN), 5x with 10µl 80% ACN in 1% acetic acid, 5x with 10µl 50% ACN in 1% acetic acid, 5x with 10µl of 30% ACN in 1% acetic acid and 5x with 10µl of 1% acetic acid. Then 10 µl of the sample were aspirated and dispensed 10 times. An additional washing step 5x with 10 µL of 1% acetic acid removed contaminations. Peptides were eluted using 5x 5µl 50% ACN in 1% acetic acid and 5x5µl 80% ACN in 1% acetic acid. The volume of the samples was reduced to 5 µl by vacuum centrifugation and finally filled up with 15 µl 2% ACN in 0.1% acetic acid. As ZipTips had a maximal loading capacity of 2 µg, the samples had a concentration of 0.1µg/µl. Samples were stored at -80°C prior to use.

Mass spectrometric analysis

Prior to MS-analysis using LTQ-FTICR-MS (Thermo Fisher Scientific) the samples were fractionated using the nanoACQUITY UPLC (Waters) equipped with a C_{18} nanoAcquity column (100 µm x 100 mm, 1.7µm particle size).

The four technical replicates were analysed subsequently and at least five washing steps were performed between different biological samples.

MS-data were generated using the LTQ-FTICR-MS equipped with a nanoelectrospray ion source (PicoTip Emitter FS360-20-20-CE-20-C12, New Objective). After a first survey scan in the FTICR (r= 50,000) MS2 data were recorded for the five highest mass peaks in the linear ion trap at a collision induced energy of 35%. The exclusion time was set to 30 sec and the minimal signal for

MS2 was 1000. A database search was performed using the transproteomic pipeline via the Sorcerer platform using the *P. putida* KT2440 database according to the annotation available at GenBank (Acc.-No.: AE015451.1, http://cmr.jcvi.org/tigr-scripts/CMR/GenomePage.cgi?database=gpp). Mass tolerances for protein identification by MS2 were set to 100 ppm. Up to two missed tryptic cleavages were considered, methionine oxidation was set as variable modification. The results from the database search were loaded into the Scaffold 2.0 software (Proteome Software) for further analysis. Data were loaded as one biological sample to obtain the total number of proteins detected in the entire analysis. Peptides that belong to more than one protein were excluded from the analysis. For significant identification of a protein two unique peptides (95% peptide probability), three spectra and a protein probability of 99% were required. For quantitative analysis technical replicates of each sample were loaded again as one biological sample into Scaffold 2.0. A protein was considered to be present and only included in the quantitative analysis if two unique peptides were detected in at least two technical replicates and in all samples of one group (condition to be tested, here 30°C and 10°C). Normalised spectral data were provided by Frank Schmidt and used for further analysis.

3 RESULTS AND DISCUSSION

In the present work, the stress responses to either low temperature or high sodium benzoate concentration were chosen to be examined in more detail, mainly on the transcriptome level using the *P. putida* Genome Oligonucleotide Array (see 2.1.2.1). For cold stress response a combination of three different transcriptome platforms (Progenika and Affymetrix genome microarrays and Illumina deep RNA sequencing technology) was undertaken. Furthermore metabolome and proteome data were integrated to gain a deeper insight into the stress response, to compare and validate the different transcriptome platforms in detail and additionally assess the different "omics". In the following the term Progenika and Affymetrix genome microarrays or Illumina deep RNA sequencing technology will be abbreviated with Progenika, Affymetrix or Illumina respectively.

The general mechanisms in stress response to the chosen conditions are already known. A decrease in temperature leads to reduced translational efficiency due to the formation of stable mRNA secondary structures and to hampered ribosome activity and furthermore negatively affects transport processes due to impaired membrane fluidity (Phadtare, 2004). Sodium benzoate is an organic aromatic acid that is widely used as a food preservative (Nair, 2001). Aqueous sodium benzoate enters the cell by diffusion as uncharged benzoic acid and thereby releases protons that lead to an acidification of the cytosole. As an aromate, benzoic acid can also dissolve in the cytoplasmic membrane and thereby disorganize the membrane potential, leading to a loss of lipids and proteins. In high amounts these effects can lead to cell death (Sikkema *et al.*, 1995; Comes & Beelman, 2002).

However, much remains to be discovered as the combinatorial analysis of *P. putida* wild type and stress-intolerant mutants on different "omics"-platforms should reveal new mechanisms to cope with abiotic stresses helping to gain novel knowledge and deeper insights into the stress response cascade.

3.1 Verification of Tn5 Plasposon Mutants

The Tn5 plasposon mutant library was previously generated using the strain *P. putida* KT2440 and screened for stress-intolerant, non auxotrophic mutants. The stress response to low temperature (4°C), low pH (4.5) and growth in 0.8 M urea and 45 mM sodium benzoate was determined by mRNA expression profiling. In the study by Reva *et al.* (2006) 32 genes located in 23 operons were identified to be impaired in growth when exposed to at least one of the defined stress conditions. In the present study 32 mutants affected in 25 genes that were found to be sensitive to either benzoate or cold stress were investigated in more detail and are listed in Table 3.1. A schematic overview about the experimental setup for the mutant verification is shown in Figure 3.1.

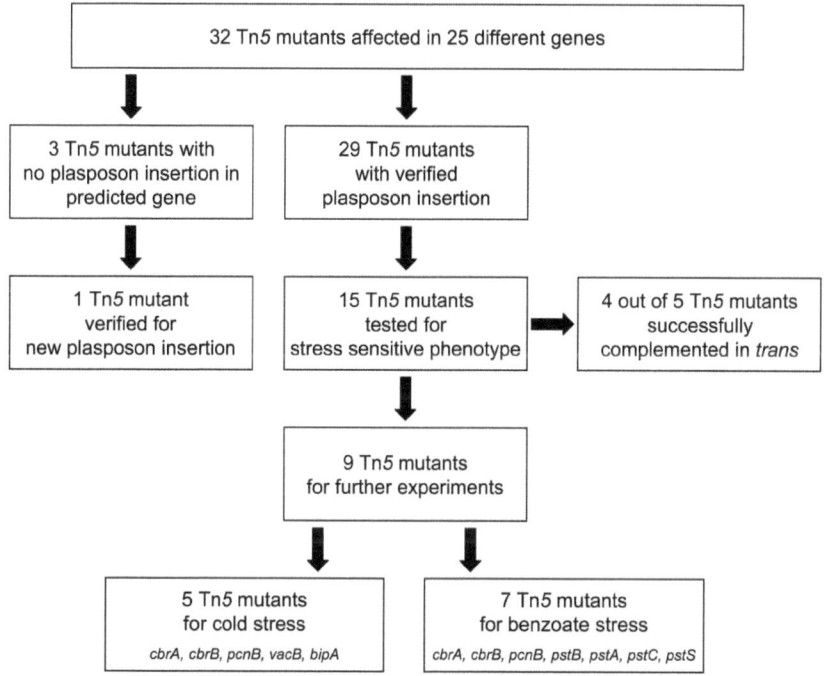

Fig. 3.1 Schematic overview of verification experiments including 32 Tn5 plasposon mutants.

Table 3.1 *Pseudomonas putida* KT2440 Tn5 plasposon mutants sensitive to cold or benzoate.

ORF with Tn5 plasposon insertion	mutant name	insert position within ORF	stress sensitive phenotype	gene name	encoded protein
PP0063	12E11	539	benzoate		Lipid A biosynthesis fatty acid acyltransferase
PP0368	12D11	1748	cold		Acyl-CoA dehydrogenase, putative
PP0816	6E6	273	benzoate	cyoE-2	CyoE-2 protoheme IX farnesyltransferase
PP1654	5E8	92	cold	cysM	CysM cysteine synthase
PP1910	35B1	95	cold		Conserved hypothetical protein
PP2242	6E3	1831	benzoate	fepA	FepA outer membrane ferric enterobactin receptor
PP2242	7B7	1829	benzoate	fepA	FepA outer membrane ferric enterobactin receptor
PP3820	14C3	222	cold		Group II intron-encoding maturase
PP3951	14C2	689	benzoate	pcaI	PcaI 3-oxoadipate-CoA transferase, subunit A
PP4129	26A10	840	cold	nuoL	NADH dehydrogenase I, L subunit
PP4185	38F3	772	benzoate	sucD	SucD succinyl-CoA synthetase, alpha subunit
PP4186	2E1	71	benzoate	sucC	SucC succinyl-CoA synthetase, beta subunit
PP4646	17G6	762	cold		Ferredoxin NADP reductase
PP4695	12E7	263	cold	cbrA	CbrA sensor histidine kinase
PP4696	5B3	948	cold	cbrB	CbrB response regulator
PP4697	C6F9	731	cold	pcnB	PcnB poly(A) polymerase
PP4731	5H1	394	benzoate	omlA	OmlA outer membrane lipid protein
PP4880	37E10	970	cold	vacB	VacB RNase R
PP4880	38C12	1482	cold	vacB	VacB RNase R
PP4941	21B5	186	benzoate		Conserved hypothetical protein
PP4982	36D4	382	benzoate		Conserved hypothetical protein
PP5044	21A2	845	cold	bipA	BipA GTP-binding protein
PP5044	22C7	358	cold	bipA	BipA GTP-binding protein
PP5044	8H9	505	cold	bipA	BipA GTP-binding protein
PP5044	38D12	1110	cold	bipA	BipA GTP-binding protein
PP5044	7G1	1032	cold	bipA	BipA GTP-binding protein
PP5322	26A3	413	cold		CBS domain protein
PP5322	27F1	412	cold		CBS domain protein
PP5326	4C7	561	benzoate	pstB	PstB-PstA-PstC-PstS phosphate ABC transporter operon
PP5327	31B4	262	benzoate	pstA	PstB-PstA-PstC-PstS phosphate ABC transporter operon
PP5328	1H7	1310	benzoate	pstC	PstB-PstA-PstC-PstS phosphate ABC transporter operon
PP5329	2F11	596	benzoate	pstS	PstB-PstA-PstC-PstS phosphate ABC transporter operon

3.1.1 Determination of plasposon insertion sites via combinatorial colony hot-start PCR

In this study, geno- and phenotype of stress-intolerant mutants identified under either cold or sodium benzoate stress were verified and growth characteristics were analyzed in more detail.

First, 32 mutants with a total of 25 different affected genes were examined for plasposon insertions using a combinatorial colony hot-start PCR. Therefore, candidate mutants were freshly grown in LB-Gm medium at 30°C overnight for use as a template for colony hot-start PCR. A specific combination of three primers, a primer pair flanking the plasposon insertion and one primer complementary to the gentamicin resistance cassette of the plasposon, was used in the PCR to identify the correct insertion site and furthermore the orientation of the plasposon. In case of correct plasposon insertion as predicted by Reva *et al.* (2006) the resulting PCR products could be calculated beforehand and were of expected size, in the other case plasposon insertion was identified by the Y-linker method and subsequent sequencing.

The mutant collection had been stored in 96-well plates for long-term storage at -80°C. Candidate mutants were transferred into fresh LB-Gm medium, incubated overnight at 30°C and were then stored separately in cryotubes as fresh glycerol-stocks at -80°C. Some PCR reactions revealed cross-contaminations in the original stock cultures from the 96-well plates that might have been occurred either during the initial inoculation of plates or during the preparation of single stock cultures. Cross-contaminations were visible in the PCR reactions when products generated on mutant DNA with the gene specific primers flanking the plasposon insertion showed a PCR product that had the same size as control PCR product generated on wild type DNA using the same primers. Contaminated mutant strains were purified by spreading single colonies on LB-Gm agar plates. Single colonies were then picked and examined for purity and correct plasposon insertion site. Positive single colonies were used to prepare a new stock culture.

In total 29 of the 32 candidate mutants (see Table 3.1) could be verified via combinatorial colony hot-start PCR to have a plasposon insertion at the predicted site, three mutants did not exhibit the predicted plasposon insertion. Out of the three mutants, the two initial mutants, *fepA*::Tn5:1829 and *bipA*::Tn5:1110, were excluded from further analysis as other mutants with a plasposon insertion in the same gene but at different insertion sites existed and thus could still be examined for their stress responsive phenotype. An unexpected insertion site was verified for mutant 12E11, as described in 3.1.2.

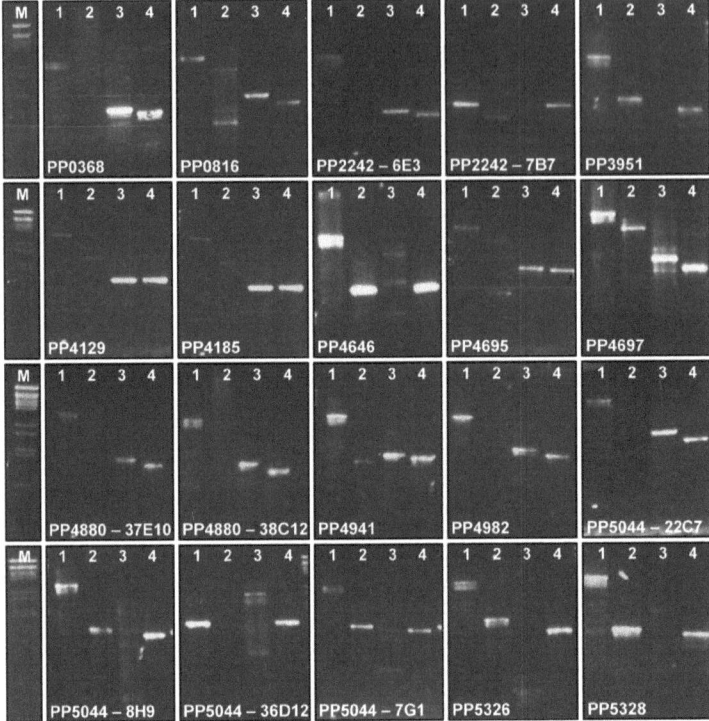

Fig. 3.2 Electrophoresis of combinatorial colony PCR products of selected KT2440 plasposon mutants obtained with primer combinations forward and reverse primer flanking the target gene (lane 1), forward and trans1 primer (lane 2) or reverse and trans 1 primer (lane 3) using overnight cultures as template. PCR products obtained with forward and reverse primer flanking the respective gene using wild type DNA was applied as positive control (lane 4). PP2242-7B7 and PP5044-36D12 show no pTnMod-OGm insertion at the predicted site, all other PCR products are specific for pTnMod-OGm insertion of expected size. M: λ DNA-BstEII digest.

3.1.2 Identification of the correct plasposon insertion site in mutant 12E11

The combinatorial colony hot-start PCR did not show the predicted plasposon insertion in gene PP0063 (see Figure 3.3, A). Since the depicted mutant was still gentamicin resistant, it was more likely that the predicted insertion site was incorrect, rather than a loss of the plasposon.

For this reason, other stock cultures remained from the initial screening which was supposed to contain PP0063::Tn5 mutants were selected and examined for plasposon insertion by designing a primer pair specific for the gentamicin cassette in the plasposon. In total 10 different stocks (in the following named A-J) were analysed, of which nine could be verified to carry a plasposon insertion by generating a PCR product with the gentamicin specific primers (see Figure 3.3, B). In parallel, Southern blot hybridization was performed with two DNA probes complementary to sequences specific for gene PP0063 or the gentamicin resistance cassette, respectively (see Figure 3.3, C). Genomic DNA was digested with either the restriction enzymes PstI or BamHI that do not cut within the plasposon sequence. Due to restriction sites in the region around PP0063, the enzymes PstI and BamHI produce probe specific fragments of 1266 bp and 2162 bp, respectively, from wild type DNA. In case of a plasposon insertion in PP0063 the fragment would be approximately 1.6 kb larger. The hybridization results demonstrated that in nine out of ten cases the plasposon insertion site was incorrect and for one mutant strain a plasposon insertion could not be detected (see above). The Southern Blot revealed three different insertion sites, two (in the following named A and B) of which were identified via sequencing of Y-linker PCR products. Hereby, genomic DNA derived from mutants A and B as displayed in Figure 3.3, C, was digested with NlaIII and Sau3AI and subsequently ligated with the Y-linker. The resulting PCR products (see Figure 3.3, D) were sent for sequencing using both Tn2 primers and Y-primers. The obtained sequences were aligned against the *P. putida* KT2440 genome using the Blast algorithm (http://blast.ncbi.nlm.nih.gov/Blast.cgi). Thus the genes *kgdA* (PP4189; mutant A) and *cyoE-2* (PP0816; mutant B) could be identified as the mutated loci, with plasposon insertion sites at position 2337 bp and 749 bp within the genes, respectively.

The gene *kgdA* is organized in an operon together with *kgdB*. These genes encode the 2-oxoglutarate dehydrogenase component E1 and E2 which is an important enzyme in the citric acid cycle catalyzing the first step in the conversion of 2-oxoglutarate to succinyl-CoA. Two other mutants verified in this study, with insertions in *sucC* and *sucD*, the subunits of succinyl-CoA synthetase, also belong to the enzyme complex of the citric acid cycle. These enzymes convert succinate into succinyl-CoA and vice versa. The three mutants had concordant phenotypes: all were sensitive to high concentrations of benzoate which is unsurprising as utilization of benzoate funnels directly into the citric acid cycle. The second identified mutant had the plasposon insertion in *cyoE-*

2. A comparable mutant was already present among the verified mutants (mutant 6E6, see Table 3.1).

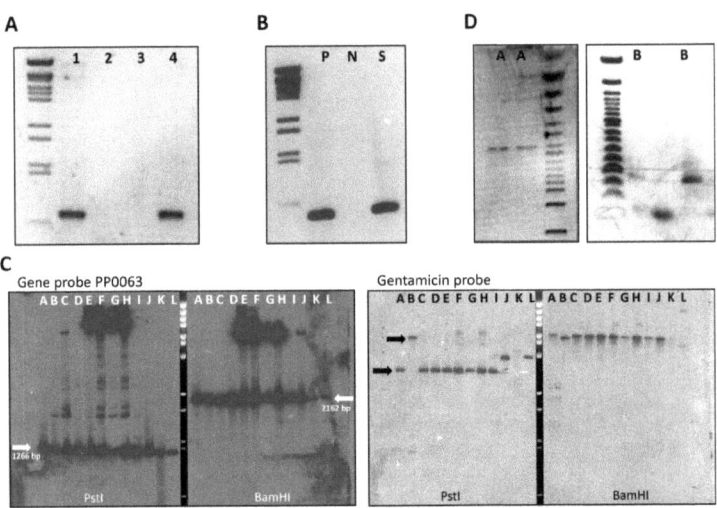

Fig. 3.3 **A)** Electrophoresis of combinatorial colony PCR products of plasposon mutant 12E11 obtained with primer combinations as described for Fig. 3.1 showing no pTnMod-OGm insertion in gene PP0063 (lane 1). λ DNA-BstEII digest, left lane. **B)** Electrophoresis of PCR products obtained with primer genta-f and genta-r specific for gentamicin resistance cassette of pTnMod-OGm using DNA from mutant PP5044-22C7 as positive control (lane P), from wild type as negative control (lane N) and from mutant 12E11 (lane S) proving pTnMod-OGm insertion in mutant 12E11. λ DNA-BstEII digest was used as molecular weight marker, left lane. **C)** Southern blot with DNA from predicted PP0063::Tn5 mutants (A-J) hybridized with DNA probes specific for gene PP0063 (left) and gentamicin resistance cassette of pTnMod-OGm (right). Genomic DNA digested with restriction enzymes PstI and BamHI resulting in fragments of 1266 bp and 2162 bp in size respectively, were detected by the PP0063 specific probe (left). Southern blot hybridized with DNA probe specific for gentamicin resistance cassette showed plasposon insertion at three different sites of the *P. putida* KT2440 genome (A, B, J). Black errors indicate candidate mutants for Y-linker verification. λ DNA-BstEII digest was used as size marker. A-J: Mutants, K: *P. putida* KT2440 wild type (negative control), L: PP5044-22C7 (positive control). **D)** Electrophoresis of PCR products obtained with y-primer and tn2 primer using DNA from candidate mutants A and B that were then sent for sequencing. 100 bp DNA ladder (right picture, left lane) and 50 bp DNA ladder (left picture, right lane) were used as molecular weight marker.

3.1.3 Phenotypic verification of stress sensitive mutants

After verification of the respective plasposon insertion, 15 were selected for further analysis. The excluded mutants had either no predicted functional annotation for the affected gene or did not show comparable strong growth inhibition in the initial screening under high sodium benzoate concentrations (Reva et al., 2006).

For the phenotypic verification, cultures were started in LB broth and after overnight incubation at 26°C transferred in 20 mL fresh medium. For this, cells of overnight cultures were centrifuged (6000 g, 3 min) and cell pellet was resuspended in respective medium to deplete LB medium that might lead to enhanced growth due to residual carbon sources. All mutants were tested under standard (ABC medium + 15 mM sodium benzoate) and both stress conditions.

Benzoate stress: ABC medium supplemented with 45 mM sodium benzoate, 26°C
Cold stress: Standard medium, growth temperature of 5°C

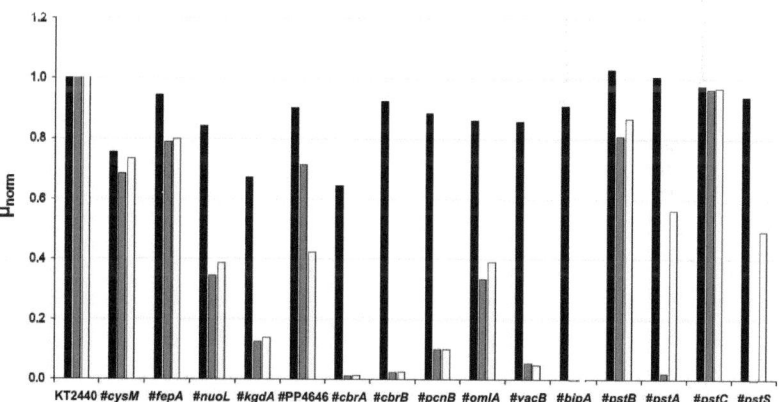

Fig. 3.4 Normalized growth rates μ_{norm} of selected plasposon mutants in comparison to wild type growth after 12 and 32 hours when grown on 15 mM or 45 mM sodium benzoate at 26°C and on 15 mM sodium benzoate at 5°C respectively. Black: standard condition (ABC minimal medium + 15 mM sodium benzoate, 26°C), dark grey: high sodium benzoate concentration, light grey: low temperature.

Growth was observed for two parallel cultures by measuring optical density at different time points for up to 12 and 32 hours when grown at 26°C and 5°C respectively. Growth was determined by calculating the growth coefficient that was then normalized to wild type growth. The growth coefficient was determined by transforming the mean value percentage for optical density for each time point by natural log (ln) and calculating the slope of the growth curve for the time period of interest: $\mu = [\ln(OD_1)-\ln(OD_0)] / (t_1-t_0)$; $\mu_{norm} = \mu_{mutant} / \mu_{wt}$.

Mutant growth under standard condition was comparable to the wild type; only three mutants (*cysM*::Tn5, *kgdA*::Tn5 and *cbrA*::Tn5) exhibited reduced growth rates of less than 0.8. Two mutants, *cysM*::Tn5 and *pstC*::Tn5, could not be verified to be stress sensitive under either of the two stress conditions whereby the mutant *cysM*::Tn5 displayed on overall reduced growth rate of 0.8. Another two mutants, *fepA*::Tn5:1831 and *pstB*::Tn5, only showed a slight decrease in growth rate when grown under stress conditions. All other mutants were strongly impaired in growth under one of the tested stress conditions and growth rate was less than 0.4. Interestingly, most stress sensitive mutants were similarly repressed under both cold and benzoate stress conditions. A distinct phenotype in response to either one of the tested conditions could be observed for just three mutants, PP4646::Tn5, *pstA*::Tn5 and *pstS*::Tn5. PP4646::Tn5 had a reduced growth rate of 0.7 when grown on 45 mM benzoate and 0.4 when grown at 5°C. For the two mutants, *pstA*::Tn5 and *pstS*::Tn5, the difference was even more significant. Growth on 45 mM benzoate was almost diminished whereas the growth rate at 5°C was 0.5.

In summary, for four out of the 15 mutants a strong stress sensitive phenotype could not be observed. Six mutants were extremely impaired in growth under both tested stress conditions with a reduced growth rate less than 0.15. The mutants affected in the four *pstBACS* genes encoding the different subunits of an ABC phosphate transporter showed a discordant phenotype. Growth of *pstB*::Tn5 and *pstC*::Tn5 were not considerably repressed by either of the stress conditions, in contrast to *pstA*::Tn5 and *pstS*::Tn5 that displayed a strong benzoate sensitive phenotype.

Only the mutants with the strongest phenotype were selected for more detailed analysis. Mutants *cbrA*::Tn5 and *cbrB*::Tn5 (CbrAB two-component system) together with mutant *pcnB*::Tn5 mutant (adjacent gene encoding for a poly(A) polymerase), were examined under both benzoate and cold stress. In the initial screening the *cbrAB*::Tn5 and *pcnB*::Tn5 mutants had only been found to be cold but not benzoate sensitive during growth under either of the stress conditions. However, transcriptome data also indicated a role for *cbrAB* in the benzoate stress response as this operon was 2.5-fold down-regulated in stationary cells grown on 45 mM benzoate (Reva *et al.*, 2006). CbrAB is described to play a paramount role in central metabolism by regulating a healthy C:N balance

(Nishijyo et al., 2001; Zhang & Rainey, 2008). The PcnB poly(A) polymerase governs polyadenylation at the 3' end of mRNA and thus exerts major impact on mRNA fate (O'Hara et al., 1995; Sarkar, 1996, 1997). The three genes show the same phenotype in response to the tested stress conditions and are located identically arranged in all *Pseudomonas* genomes sequenced so far, suggesting a functional interplay.

The two mutants *vacB*::Tn5 and *bipA*::Tn5 affected in genes encoding ribonuclease R and GTP-binding protein, respectively, were subsequently only examined under cold stress due to their genetic background. The VacB RNAse R is involved in posttranslational regulation of mRNA stability, the GTP-binding protein BipA is associated with ribosome structure and function. Both features are important defence or repair mechanisms in response to cold shock (Erova et al., 2008; Krishnan & Flower, 2008).

Though only two out of the four *pstBACS*::Tn5 mutants showed a benzoate sensitive phenotype, all were included in the following benzoate stress experiments.

3.1.4 Complementation in *trans*

Mutants affected in genes that were either organized in operons or were found to be multiply affected in the same gene but with different plasposon insertion sites (*vacB*::Tn5 and *bipA*::Tn5) in the initial screening, were not complemented in *trans*. For the excluded mutants a consistent phenotype was regarded as proof to be caused by the plasposon insertion and not by secondary mutation events. However, the remaining five mutants apart from *fepA*::Tn5:1831, for which stress sensitivity could not be verified, were complemented in *trans* as described here.

The observed growth at 5°C in ABC minimal medium supplemented with 15 mM sodium benzoate was overall very poor, even the wild type grew only with a mean growth rate of 0.1 h^{-1}. The reversion of the phenotype due to the complementation could not be tested under this condition. Secondly, benzoate is not an optimal carbon source on and might implement an enhanced stress effect when grown on benzoate at low temperature. Therefore, the cold stress experiments were thereupon performed in M9 minimal medium supplemented with 15 mM succinate, a preferred carbon source, at 10°C to maintain more robust growth conditions.

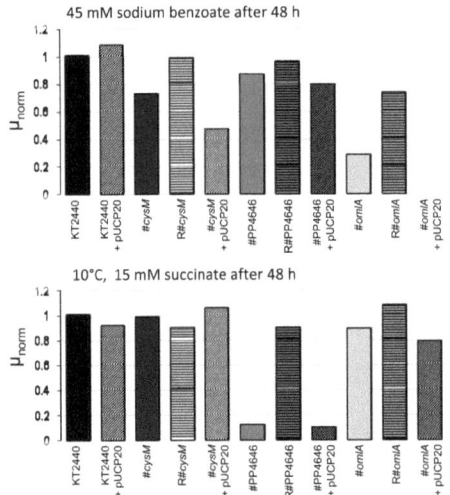
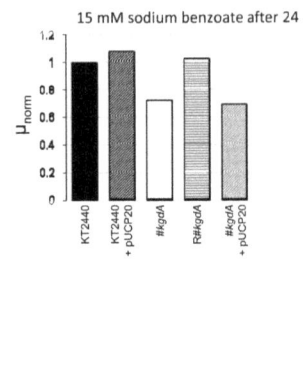

Fig. 3.5 Normalized growth rate of strains used for complementation experiments under 45 mM and 15 mM sodium benzoate at 26°C or 15 mM succinate at 10°C. Black: *P. putida* KT2440 wild type, dark grey: *cysM* mutant (PP1654), grey: mutant PP4646, light grey: *omlA* mutant (PP4731), white: *kgdA* mutant (PP4189). Plain bars represent wild type and plasposon mutants, respectively, diagonal dashed bars represent strains containing empty pUCP20 vector as negative control, horizontal dashed bars represent revertants containing pUCP20 vector with the respective gene and associated promoter region.

Complementation of the cold sensitive mutants was therefore tested under the new defined conditions (see Figure 3.5).

Mutants that were successfully cloned with the pUCP20 vector containing the respective gene sequence were tested for recovery of the wild type phenotype. Wild type phenotype could be restored for mutants PP4646::Tn5 and *omlA*::Tn5 under both tested stress conditions, the effect was most significant for mutant PP4646::Tn5 when grown at 10°C, which verifies the distinct stress sensitive phenotype for benzoate and cold stress (see Figure 3.4). For mutant *omlA*::Tn5, the effect was more significant under high concentrations of benzoate, indicating that the previously observed cold sensitivity was based on the growth medium with benzoate as sole carbon source. The same could be shown for mutant *cysM*::Tn5 that could be complemented for benzoate stress whereas no growth difference between wild type, mutant and complemented one could be observed for cold stress. Mutant *kgdA*::Tn5 could merely be complemented for growth on 15 mM benzoate but not for the defined stress conditions. Mutant *nuoL*::Tn5 could not be complemented under any condition.

The complementation experiments indicated, as seen for mutant PP4646::Tn5 and *omlA*::Tn5, that using benzoate medium for cold stress experiments is an additional stress factor. Therefore, mutants selected for further cold stress experiments were verified to be cold sensitive when grown on succinate, a more preferential carbon source than benzoate. Results are described in the next chapter.

Discussion

Transposon mutant libraries constitute an important approach for studying bacterial genomes. Independent studies of *Pseudomonas* transposon mutant libraries showed that, by identification of genes essential for a wide range of biological processes, researchers can gain new insights into the metabolic network and identify links of so far unpredicted relationships between different regulatory routes and growth states.

First comprehensive transposon mutant library studies in *P. aeruginosa* revealed 300 to 400 genes to be essential for growth on rich media. From these candidate essential genes the majority could be attributed to the functional classes of translation and cell division (Jacobs *et al.*, 2003; Liberati *et al.*, 2006). Since then, a number of *Pseudomonas* transposon mutant libraries have been generated and screened for various phenotypes (Overhage *et al.*, 2007; Liberati *et al.*, 2008). Though the researchers screened for different auxotrophic phenotypes and mutants affected in motility, the number of identified mutants is comparable to the study by Reva *et al.* (2006).

Recently, two Tn*5* transposon mutant libraries have been generated for *P. putida*. In strain *P. putida* CA-3 Goff *et al.* (2009) identified two mutants to be affected in PHA (polyhydroxylalkanoate) accumulation and to overproduce lipopolysaccharides when grown at low temperatures. PHAs are biocompatible and biodegradable polymers that accumulate intracellularly under limited nitrogen or phosphorus but excessive carbon conditions. Since these polymers are of interest in the fields of biomedicine and packaging, it is relevant to better understand factors affecting PHA biosynthesis and hence to improve the process for industrial applications.

The study of *P. putida* KT2440 by Molina-Henares *et al.* (2010) revealed 48 auxotrophic mutants to be impaired in growth on minimal medium with glucose as sole carbon source. Among these, many were found to be involved in certain amino acid biosynthesis pathways. The metabolic complementation of L-arginine, L-leucine, L-proline and L-cysteine with the corresponding D-amino acids suggests a number of amino acid racemases to be encoded on the *P. putida* KT2440 genome.

None of the mutants identified by Goff *et al.* (2009) or Molina-Henares *et al.* (2010) were found in this study. However, they could also identify mutants to be affected in genes that were so far not attributed to the described processes. Hence, transposon mutant libraries are a feasible tool for the rapid identification of genes that were previously not associated with the tested biological processes and might be of importance for future biotechnological applications.

However, though the affected genes causing stress sensitive phenotypes were previously identified (Reva et al., 2006), the verification of the correct genotype and consequential phenotype using more standardized growth conditions was necessary. Out of the 32 mutants predicted to be either cold or benzoate sensitive, the predicted plasposon insertion site was incorrect for just three mutants. Furthermore, for another three out of the 15 selected mutants tested under more standardized growth conditions, a stress sensitive phenotype could not be observed. Though screening of 96-well plates of transposon mutant libraries for striking phenotypes provides high-throughput, the detailed verification of selected mutants for correct genotype and consequential phenotype is an important feature.

However, the task of elucidating the global stress response caused by these mutations remains. For this reason, cold sensitive mutants were selected for transcriptome and metabolome profiling in response to cold shock stress.

3.2 Growth Experiments under Stress Conditions

Nearly a decade ago systems biology emerged as a prominent new field in bioscience research. This interdisciplinary area is an approach to understand biological systems as a whole, rather than by their constituents alone. The combination of experimental gained knowledge from different biological platforms as provided by transcriptome, metabolome and proteome together with mathematical models and model-driven predictions provides new insights into the organization of biological systems and by *in-silico* predictions might even help to design experiments in a more efficient way. To integrate results from different platforms in a feasible way and to be able to make reliable predictions, an important feature is the standardization of experimental setups.

This project was a part of the PSYSMO consortium (http://www.psysmo.org/), one of eleven projects funded by the European transnational funding and research initiative on "Systems Biology of Microorganisms (SysMO). In this context, growth conditions were standardized to ensure comparable results among different laboratories involved in this consortium. Standard conditions were defined as follows:

- M9 minimal medium (Miller, 1972), pH 6.8
- Supplemented with 15 mM succinate as sole carbon source
- Additionally supplemented with $MgSO_4$, $FeSO_4$ and trace elements to provide efficient growth.
- Temperature: 30°C

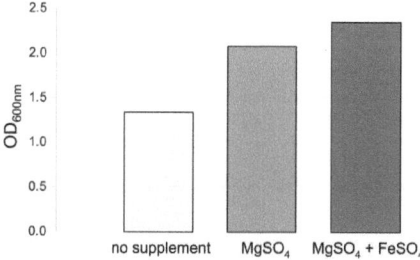

Fig. 3.6 Final optical density (600_{nm}) for *P. putida* KT2440 wild type when grown on ABC minimal medium for 24 h containing 15 mM sodium benzoate as sole carbon source (light grey), supplemented with 2 mM $MgSO_4$ (grey) or 2 mM $MgSO_4$ and 10 µM $FeSO_4$ (dark grey), respectively.

Supplementation of MgSO$_4$ and FeSO$_4$ resulted in significantly higher cell density, as shown in Figure 3.6. For this reason, the medium for further stress assays was changed from ABC minimal to M9 minimal medium and temperature from 26°C to 30°C. The growth experiments were still performed in 100 mL flasks with a culture volume of 20 mL. For the comparable analysis of the "omics"-platforms, strains were cultured in the BioFlo 110 fermenter with a volume of 1.5 L to ensure more standardized conditions.

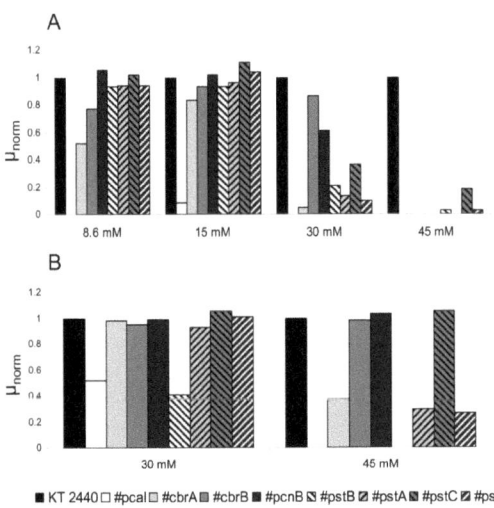

Fig. 3.7 Normalized grows rates of benzoate sensitive mutants in comparison to growth of *P. putida* KT2440 wild type under increasing sodium benzoate concentrations. **A)** Normalized grows rate of early growth phase from 0 to 8 hours. **B)** Normalized growth rate from 8 to 24 hours.

3.2.1 Growth Characteristics under Increasing Sodium Benzoate Concentrations

Benzoate sensitive mutants described in the previous chapter were selected for the benzoate stress assay under increasing sodium benzoate concentrations. For this, strains were initially incubated in LB broth for eight hours and then transferred in M9 minimal medium supplemented with 8.6 mM sodium benzoate and incubated overnight. The intermediate culture was used to ensure that the strains were already adapted to benzoate as carbon source and any observed stress response was only due to increased benzoate concentrations. Growth was monitored in four different benzoate concentrations ranging from 8.6 to 45 mM for up to 24 hours (see Figure 3.7). Since mutants were still capable of growing on benzoate as sole carbon source, benzoate sensitivity was mainly displayed in a prolonged lag-phase. Once adapted to the benzoate medium, mutants grew sufficiently with a final cell density comparable to the wild type after 48 hours. For this reason, separate growth rates were determined for early (0-8 hrs) and late (8-24 hrs) growth phase and normalized to the wild type.

The mutants exhibited no significant sensitivity to benzoate up to a concentration of 15 mM. Only the *cbrA*::Tn*5* and *cbrB*::Tn*5* mutants showed a slightly decreased growth rate in the first eight hours when grown on 8.6 mM benzoate ($\mu_{wt} = 0.4$ h^{-1}). Nevertheless, the growth rate of all mutants from 8 to 24 hours was comparable to the wild type. An increased sodium benzoate concentration of 30 mM, however, had a marked impact. When comparing early and late growth phase under 30 mM benzoate, all mutants showed decreased growth in the early growth phase ($\mu_{wt} = 0.2$ h^{-1}), but were able to grow similar to the wild type once they have been adapted to the high concentration of benzoate. Only the *pstB*::Tn*5* mutant was overall impaired in growth. The effect was even more striking when grown on 45 mM benzoate. During the first eight hours ($\mu_{wt} = 0.1$ h^{-1}), mutants displayed very little growth. Only the *cbrB*::Tn*5*, *pcnB*::Tn*5* and *pstC*::Tn*5* mutants were able to grow comparable to the wild type, once they have been adapted ($\mu_{wt} = 0.2$ h^{-1}). The normalized growth rates of the remaining mutants were less than 0.4.

In addition to the seven mutants from the initial screening, the *pcaI*::Tn*5* mutant was included into the assays for sodium benzoate stress. This mutant exhibited a very long lag-phase as demonstrated by nearly no visible growth under any benzoate concentration in the first eight hours, but was then able to grow on benzoate. This result will be discussed in more detail in chapter 3.3.

3.2.2 Comparison of growth under cold adaptation and cold shock

Initially, the cold sensitive phenotypes of the selected mutants were verified for growth in the newly defined standard medium. To ensure that the observed phenotype resulted only from growth at a low temperature, strains were, similarly to the benzoate stress assay, incubated in an intermediate culture (here M9 + 15 mM succinate) so that mutants were already adapted to the medium when transferred to the lower temperature.

When grown constantly at 10°C, mutants showed a reduced normalized growth rate of at least 0.7 relative to the wild type ($\mu_{wt} = 0.1\ h^{-1}$). Solely the $cbrA$::Tn5 mutant did not display a cold sensitive phenotype, indicating that the observed phenotype in the initial screening was caused by the benzoate medium, rather than the decreased temperature, since the cbrA::Tn5 mutant had already been demonstrated to be benzoate sensitive (see chapter 3.2.1, Figure 3.7).

In parallel, mutants were also monitored for their response to cold shock. For this, strains were grown at 30°C to mid-exponential phase (OD_{600nm} ~0.8) and immediately transferred to 10°C. Growth rates were determined for 30°C and 10°C. When grown at 30°C, only the $cbrB$::Tn5 mutant displayed a reduced normalized growth rate of less than 0.7 ($\mu_{wt} = 0.7\ h^{-1}$). Growth rates after applying the cold shock were generally reduced by about 85% ($\mu_{wt} = 0.1\ h^{-1}$), but compared to the wild type, only three mutants, $pcnB$::Tn5, $vacB$::Tn5 and $bipA$::Tn5, showed a significantly reduced normalized growth rate of less than 0.6. The $cbrB$::Tn5 mutant on the contrary exhibited a generally impaired growth independent of the cold shock, but was still declared as cold sensitive as the growth rate was significantly more reduced when grown constantly at 10°C.

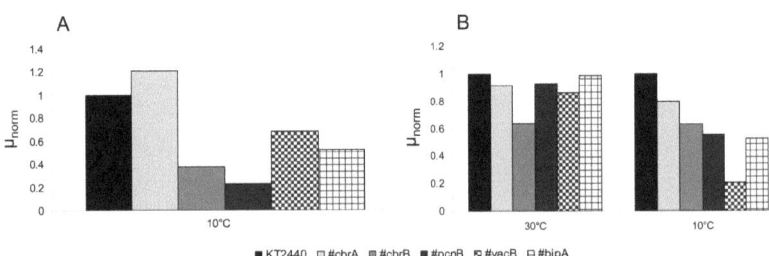

Fig. 3.8 Normalized growth rates of cold sensitive mutants in comparison to growth of *P. putida* KT2440 wild type cultivated in M9 minimal medium supplemented with 15 mM succinate. **A)** Cold adaptation, **B)** Cold shock.

RESULTS AND DISCUSSION

To determine the setup for future experiments such as transcriptome, metabolome and proteome analysis, the gene expression profile of the wild type in response to either cold adaptation or cold shock was examined.

The gene expression profile was determined by cDNA hybridization of Progenika microarrays. The wild type strain was cultivated under standardized growth conditions in the BioFlo 110 fermenter. For the reference sample, cells were grown to mid-exponential phase ($OD_{600nm} \sim 0.8$) at 30°C and harvested for subsequent RNA preparation. After sampling, the temperature was decreased from 30°C to 10°C for one hour, simulating a cold shock. When the medium reached 10°C, cells were immediately harvested for a direct cold shock sample and again four hours later for an adapted cold shock response. For the cold adaptation experiment, cells were grown constantly at 10°C, again to mid-exponential growth phase. After harvesting, samples were subsequently processed for RNA extraction, with three aliquots of cells used in parallel to produce technical replicates. The reference sample was used for all three comparisons. For each condition, direct and adapted cold shock response and cold adaptation, four technical microarray replicates were used.

After statistical evaluation of the microarray data, genes were considered to be significantly differentially regulated, if p-value ≤ 0.05 and FDR ≤ 0.1 were obtained for the calculated fold-change of expression with Progenika microarrays. The comparison of gene expression induced by cold adaptation processes or cold shock response revealed a significant higher number of differentially regulated genes in response to cold shock. The cold adaptation profile revealed 64 genes to be significantly expressed, of which only 36 showed a differential expression greater than two-fold. In contrast, 628 genes were detected to be significantly expressed due to cold shock (89 up- and 172 down-regulated more than two-fold). The transcriptome analysis of the cold shock sample, taken four hours after the medium reached 10°C already revealed an adaptation process on the transcriptional level. Less genes (439) were found to be significantly expressed at that time-point of which 52 were up and 58 down-regulated more than two-fold. The results are discussed in more detail in section 3.6.

Based on these results and as there were no definite suggestions that mutants are more sensitive in response to one of the tested conditions (Figure 3.8 and Figure 3.9), the cold shock setup was chosen for future experiments. The cold shock experiments were designed as described above, but cells were harvested after a shorter adaptation phase of only two hours. Though gene expression was more dynamic directly after the temperature downshift, differences on the protein level would not appear within a few minutes. Therefore, as proteome samples should be included in future experiments, a shorter adaptation time of only two hours was implemented to ensure detectable differences in protein amounts.

A further reason for this cold shock experimental setup was that for the analysis of the wild type the two comparative samples for 30°C and 10°C were derived from the very same culture, an important feature for the comparability of the results.

The growth for wild type and mutants was furthermore monitored in the standardized conditions in the BioFlo 110 fermenter (see chapter 2.4). Growth rates were determined for exponential growth at 30°C and after temperature downshift to 10°C up to the time point of sampling.

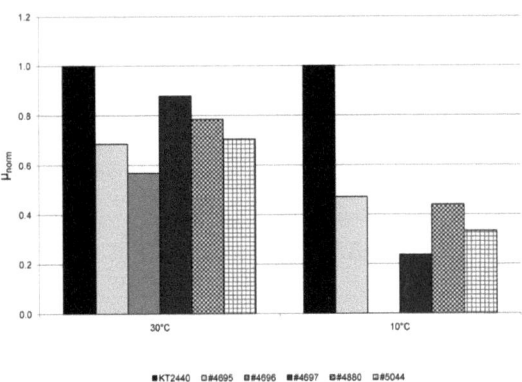

Fig. 3.9 Normalized growth rates of cold sensitive mutants compared P. putida KT2440 wild type in M9 minimal medium supplemented with 15 mM succinate when cultivated in the BioFlo 110 fermenter. When grown to mid-exponential phase (OD_{600nm} ~ 0.8) at 30°C, temperature was decreased to 10°C.

In contrast to the previous growth experiments performed in culture flasks, the growth rate under standard conditions (30°C) was reduced by more than 10% for the pcnB::Tn5 mutant and more than 20% for the remaining mutants (μ_{wt} = 0.9 h^{-1}). After temperature downshift, the growth rates of all five mutants were reduced by more than 50% (μ_{wt} = 0.1 h^{-1}). The cbrB::Tn5 mutant however, did not display any growth in the first two hours after temperature downshift under the standardized conditions.

RESULTS AND DISCUSSION

Discussion

In summary, fine-tuning of the experimental setup for the verification of growth characteristics of mutants in comparison to wild type growth is a very important feature. Large fluctuations in growth behaviour seen in this study could be attributed to the experimental setup. Though the majority of stress sensitive phenotypes as observed in the initial screening could be verified, discrepancies are explainable by the following points:

The screening in 96-well plates did not provide constant shaking and the resulting oxygen limitation might have had a strong influence on the observed growth deficiencies.

With a culture volume of only 200 µL, side effects such as evaporation could also have had a significant impact on growth behaviour.

The composition of the growth medium is an important feature. Though sodium benzoate can be easily utilized by *P. putida* KT2440, it is not the preferred carbon source compared to succinate (Aranda-Olmedo *et al.*, 2006) and thus can also influence the growth at low temperatures. Furthermore, growth on solid agar as used in the initial screening and in liquid medium is hardly comparable.

Supplementation of trace elements provides sufficient growth under standard conditions (see Figure 3.6) and therefore the absence of trace elements as in the initial screening might have been an additional stress factor under the tested stress conditions in the previous screening.

The growth conditions for the cold shock assay were standardized by growing the cells in a fermenter with constant agitation, aeration and pH. Growth was furthermore not disrupted by sampling as this is the case for growth experiments in culture flasks when cultures are stopped for measuring cell density. Thus, the cold sensitive phenotype of the selected mutants is reliable as the error rates in culture flasks is higher and therefore might explain the inconsistent phenotypes. Still, the *cbrA*::Tn5 and *cbrB*::Tn5 mutants affected in the two-component system seem to be only slightly affected.

3.3 *Pseudomonas putida pcaIJ* Plasposon Rescue

The *pcaI-pcaJ* operon of *Pseudomonas putida* KT2440 encodes the subunits of the enzyme β-ketoadipate:succinyl-CoA transferase that catalyzes the penultimate step for the dissimilation of benzoate into tricarboxylic acid intermediates. A KT2440 mutant that had a plasposon insertion close to the end of *pcaI* was investigated for its capability to grow on benzoate. In conjunction with positive growth on benzoate as sole carbon source, the hypothesis postulated by Reva *et al.* (2006) that so far uncharacterized paralogs of PcaIJ might bypass the blocked β-ketoadipate pathway by compensating for the truncated *pcaIJ* operon, was under examination.

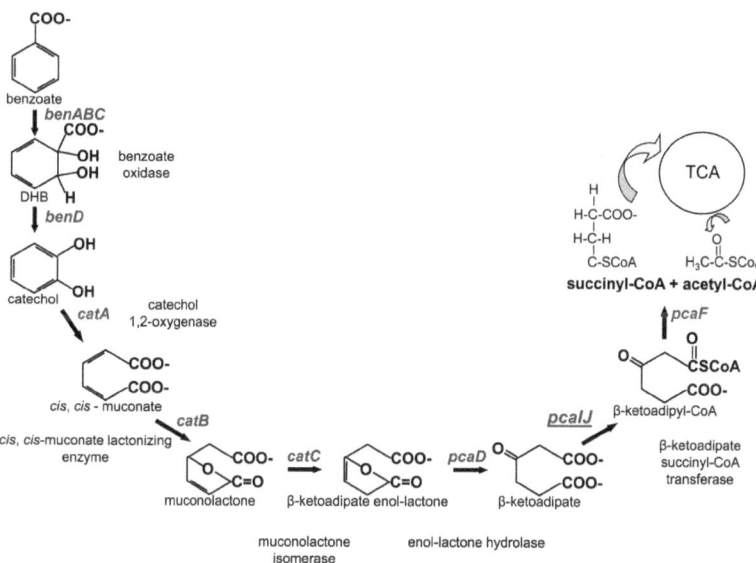

Fig. 3.10 The catechol branch of the ortho-cleavage β-ketoadipate pathway for the degradation of benzoate.

3.3.1 Wild type KT2440 outcompeted a 10^6 fold excess of an isogenic *pcaIJ* plasposon mutant

The original stock cultures of the *P. putida* KT2440 *pcaI* plasposon mutant were negative for the *pcaIJ* wild type sequence by combinatorial PCR. Growth of subcultures in LB broth and subsequent plating on LB or minimal medium with benzoate as sole carbon source, however, revealed that the stock cultures were contaminated with wild type KT2440 bacteria at a frequency of $(6 \pm 2) \times 10^{-7}$ of total cfu.

The original stock culture of the *pcaI* mutant was then tested for its ability to grow in a chemostat with benzoate as sole carbon source. An aliquot of the frozen stock was first propagated for 8 h in 5 ml LB broth and then transferred into 25 ml M9 plus 8.6 mM benzoate for overnight culturing. A 1.5 L reactor with 45 mM benzoate in M9 medium was inoculated to have $OD_{600nm} = 0.02$ of bacterial suspension. Combinatorial PCR at this time point only detected the plasposon mutant (see Figure 3.11, panel P1). After a lag phase of 15 h the bacteria started to grow and by 20 h mutant and wild type bacteria were present in similar amounts in the culture (see Figure 3.3.1, panel P2). After an $OD_{600nm} = 0.6$ had been reached 2 h later, the culture was continuously fed with M9 + 45 mM benzoate at a dilution of 0.2 per h. By 110 h, wild type KT2440 bacteria dominated the population, although the mutant was still present (see Figure 3.11, panel P3). The chemostat experiment demonstrated that the *pcaIJ* proficient wild type, which initially was present at a frequency of less than 10^{-6}, which is below the detection limit of PCR reactions, had outcompeted the isogenic mutant. This result implies that the mutant did not grow on benzoate.

3.3.2 Partial deletion of plasposon insertion rescues *pcaIJ* activity

Next, single colonies of the *pcaI* plasposon mutant were tested for growth and survival in M9 medium supplemented with 45 mM benzoate. This high benzoate concentration induces stress by membrane perturbation and cytosolic acidification (Comes & Beelman, 2002) and reduced the growth of the KT2440 wild type strain by 50% (Reva *et al.*, 2006).

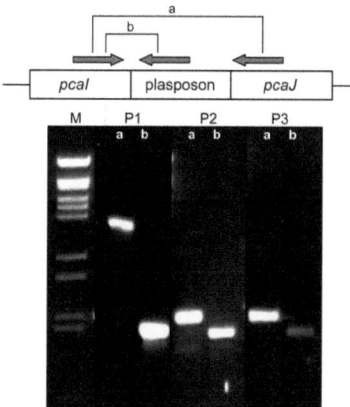

Fig. 3.11 Electrophoresis of combinatorial PCR products of the *pcaIJ* locus obtained with primer combinations F1-3951 and R1-3951 (lane a) or F1-3951 and Tn (lane b) using culture samples from the chemostat as template. P1 to P3 indicate time points of sampling at 0, 20 and 110 h of fermentation. Panel P1 lane a and b, show PCR products specific for pTnMod-OGm::KT*pcaI* of expected size. In panels P2 and P3 the PCR product in lane a is characteristic for *P. putida* KT2440 wild type. The PCR product in lane b is characteristic for pTnMod-OGm::KT*pcaI* is still present but fading. M: λ DNA-BstEII digest molecular weight marker.

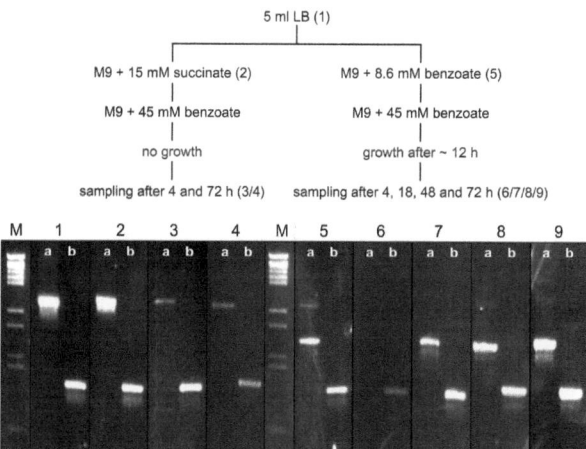

Fig. 3.12 Spontaneous partial deletions of inserted plasposon in *P. putida* KT 2440 pTnMod-OGm::KT*pcaI* during growth in benzoate. Organigram shows the two experimental strategies starting with cultures in LB broth (1) following transfer in minimal medium with either 15 mM succinate (2) or 8.6 mM benzoate (5) and subsequent exposure to stress with 45 mM benzoate (1-4 and 5-9 respectively). Gel-separated combinatorial PCR products (see Fig. 1) obtained with primer combinations F-3951 and R-3951 (a) or F-3951 and Tn (b). Panels 1 to 4 show products for pTnMod-OGm::KT*pcaI* of expected size for primer combination a and b. Subculturing of *P. putida* KT2440 pTnMod-OGm::KT*pcaI* in M9 with 8.6 mM benzoate revealed a mixture of a truncated (strong band) and the original mutant (faint band) (panel 5, lane a). Samples from the subsequent cultures in 45 mM benzoate only contained the truncated mutant (panels 6 - 9). M: λ DNA-BstEII digest molecular weight marker.

Two experimental strategies were pursued. Cultures were started in LB broth and then the bacteria were cultured with equivalent amounts of carbon source of either 15 mM succinate or 8.6 mM benzoate. Subsequently the batch cultures were exposed to stress with 45 mM benzoate (see Figure 3.12). If the *pcaI* mutant had been cultured in succinate, there was no subsequent growth in benzoate. Combinatorial PCR revealed a homogeneous population of the original plasposon mutants (see Figure 3.12). In contrast, if the bacteria had first been adapted to 8.6 mM benzoate, a concentration at which stress is not imposed upon the bacterium, subsequent bacterial growth at 45 mM benzoate was observed in six of 25 cultures. If bacteria did replicate in 45 mM benzoate, the subsequent combinatorial PCR of their *pcaIJ* locus revealed products of lower than expected size (Figure 3.12).

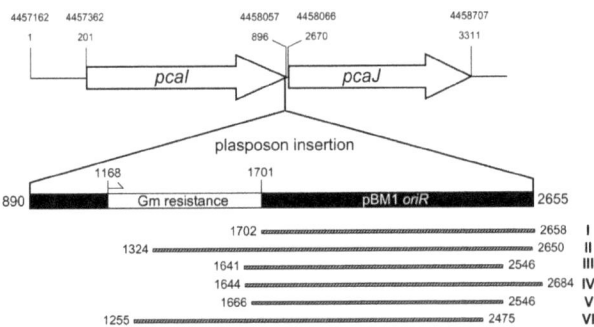

Fig. 3.13 Genetic organization of the *pcaIJ* operon with plasposon insertion in the *pcaI* gene, pTnMod-OGm::KT*pcaI*. Numbers in the uppermost row indicate the map position in the *P. putida* KT2440 genome, numbers beneath refer to nucleotide positions of the directed pTnMod-OGm::KT*pcaI* locus. Shaded lines indicate partial deletions of the plasposon insertion in the six truncated *pcaI* mutants (I-VI) uncovered by sequencing.

The sequencing of the truncated PCR products uncovered partial deletions of the plasposon insertion in all strains which had been actively growing in 45 mM benzoate (see Figure 3.13). The deletions included the origin of replication *oriR* of pMB1 (Dennis & Zylstra, 1998) and variable portions of the upstream 3-*N*-aminoglycoside acetyltransferase gene *aac3ia* that confers gentamicin resistance to the mutant (Figure 3.13). All mutants but mutant I (Figure 3.13) were susceptible to gentamicin (see Figure 3.14) implying that the resistance determinant had been inactivated.

Combinatorial RT/PCR kinetics substantiated this finding. The *pcaIJ* operon was transcribed as a single transcript in wild type KT2440. In the original mutant where the plasposon had inserted 7 nucleotides upstream of the *pcaI* stop codon, only PcaI mRNA transcript was detectable. All mutants with truncated versions of the plasposon produced wild type amounts of a fusion transcript of PcaI-AAC3IA*-PcaJ mRNA.

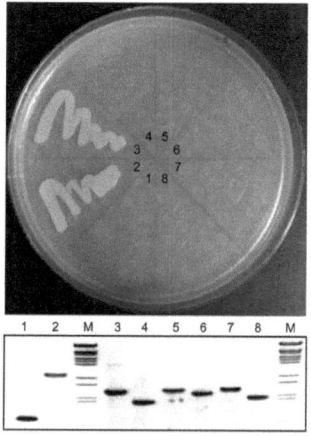

Fig. 3.14 Gentamicin resistance assay on LB plates supplemented with 40 µg/ml gentamicin for *P. putida* wild type and truncated mutants. Truncated *pcaI* mutants with a partial deletion in the gentamicin resistance cassette (II-VI) were sensitive to gentamicin. The results were supported by the same assay in LB broth supplemented with 30 µg/ml gentamicin and purity of strains was verified by colony PCR. Electrophoresis of PCR products generated with primer combination F3951 and R3951 using 1 µl of liquid culture as template shows products of expected size. 1: *P. putida* KT2440, 2: pTnMod-OGm::KT*pcaI*, 3-8 truncated *pcaI* mutants I-VI, respectively. M: λ DNA-BstEII digest molecular marker.

Discussion

The growth experiments with the *pcaI* plasposon mutant in chemostat and batch cultures consistently demonstrated that *P. putida* KT2440 can only catabolize benzoate as the sole carbon source if an intact *pcaIJ* operon is present. In other words, benzoate can only be funnelled into the tricarboxylic acid cycle via the PcaIJ and PcaF – catalyzed conversion of β-ketoadipate to acetyl-CoA and succinyl CoA. An alternative pathway that may bypass the β-ketoadipate succinyl-CoA transferase does apparently not exist in *P. putida*.

In presence of benzoate as sole carbon source the plasposon mutant could only grow if the plasposon insertion was truncated. Plasposons (Dennis & Zylstra, 1998) are modified Tn5 minitransposons (de Lorenzo *et al.*, 1990; Herrero *et al.*, 1990; de Lorenzo & Timmis, 1994) that in addition to the selection marker also encode an enterobacterial *oriR*. The origin of replication *oriR* was included into the construct because it allows for the rescue cloning and sequencing of DNA that flanks insertion sites in plasposon mutants (Dennis & Zylstra, 1998). Interestingly, only those mutants could survive, that spontaneously deleted the complete pBM1 *oriR* that functions in enterobacteria, but not in pseudomonads. G + C contents and oligonucleotide usage of the enterobacterial *oriR* are different from that of a typical *P. putida* sequence. The *Pseudomonas* host probably had to delete this foreign sequence to allow the endogeneous RNA polymerase to synthesize a functional PcaI – insert- PcaJ chimeric transcript. Tn5 minitransposons (de Lorenzo *et al.*, 1990; Herrero *et al.*, 1990; de Lorenzo & Timmis, 1994) are inherited in a stable fashion and, unlike the natural Tn5 transposon (Goldberg *et al.*, 1990), do not provoke excision, retransposition, multiple insertions, DNA rearrangements or other forms of genetic instability. The plasposons should share these features. As far as we know this is the first report that a plasposon can undergo secondary mutations under high selective pressure.

3.4 Stress Induction by Benzoate Pulse Implementation

Aromatic compounds are found as pollutants in the environment and are usually toxic to cellular systems and need to be removed from contaminated sites. An effective way to remove these compounds is microbial degradation (Parales *et al.*, 2002; Parales & Haddock, 2004) and many microorganisms isolated from these sites are capable of degrading a wide variety of aromatic compounds. Among these organisms is *P. putida* (Harwood & Parales, 1996; Wackett, 2003). Benzoate, a simple aromatic substance, is an intermediate in many biodegradative pathways, such as toluene degradation, and is therefore a good candidate for the analysis of the capability of microorganism to degrade aromatic compounds.

The goal of this study is to reveal the transcriptome profile of *P. putida* KT2440 in response to a high-dosed benzoate pulse (45 mM) over a time period of several hours. The benzoate stress experiments were performed in collaboration with Christoph Ulmer from the Department of Systems and Synthetic Biology at the HZI Braunschweig. This collaboration enabled chemostat fermentation which is the best way for standardized and reproducible growth conditions with a defined bacterial growth rate (Daran-Lapujade *et al.*, 2008).

For this, the time points to be analyzed by the Progenika microarrays during the time course of response were defined due to the outcome of RT/PCR examination of selected genes that were previously found to be up-regulated after setting a benzoate pulse.

Data acquisition and evaluation is still in progress at time of writing. Therefore, only the RT/PCR results will be discussed in the following.

Evaluation of RT/PCR Results for Time Course Experiment

The chemostat fermentation was run as follows. Pre-cultures derived from glycerol stocks were inoculated in 10 mL LB medium using 100 mL Erlenmeyer flasks for 12 hours at 30°C with constant shaking at 180 rpm. Thereafter, 1 mL of the pre-culture was used for inoculation in 50 mL M9 minimal medium (see 2.2.1) supplemented with 8.6 mM sodium benzoate as sole carbon source in 500 mL Erlenmeyer flasks and subsequent incubation for another 16 hours at 30°C. Those cultures were then used to start chemostat experiments.

The chemostat was run in a 3-1 Applikon fermenter (BioController ADI 1030, Bio Console ADI 1035, Applikon Biotechnology B.V., Netherlands) in 1.5 L working volume under the defined conditions: temperature = 30 °C, pH = 6.8 (adjusted with 20% NaOH and 10% H_2SO_4), agitation = 600 rpm with a dilution rate of 0.2 h^{-1}. The dissolved oxygen concentration pO_2 was adjusted to

50% air-saturation by mixing N_2, O_2 and compressed air using a fixed gas flow of 0.9 l min^{-1} that resulted in an aeration rate of 0.6 l l^{-1} min^{-1}. The gas flow was controlled using a mass flow meter (5850 TR, Brooks Instruments, Ede, Netherlands) and a mass flow controller (5878, Brooks Instruments).

The 1.5 L culture medium containing 8.6 mM benzoate in M9 medium was inoculated to an initial OD_{600nm} = 0.05 of bacterial suspension and was first run as a batch culture until the culture reached an OD_{600nm} of 0.8. The chemostat was then started with a dilution rate of 0.2 h^{-1}. Feed medium contained additionally 0.01% (v/v) antifoam agent (Struktol 2121, Schill & Seilacher). When steady-state was reached, here defined by the dilution rate after 25 hours, the benzoate pulse was set by increasing the benzoate concentration in the medium immediately to 45 mM.

In a first experiment, samples were taken 10 and 90 minutes after setting the pulse for subsequent RNA preparation and transcriptome analysis using Progenika microarrays. The transcriptome analysis was performed to identify candidate genes for RT/PCR that had been induced immediately after the pulse or after an adaptation phase of 90 minutes in comparison to the steady-state condition. In the second experiment, the time course of stress response was monitored in detail by performing RT/PCR using primers specific for the selected candidate genes. The stress response was examined in a period from five to 210 minutes. According to the outcome of the RT/PCR results the time points of sampling for the benzoate pulse experiment using microarrays were defined.

Table 3.2 Candidate genes that were induced after the benzoate pulse (45 mM) and therefore selected for RT/PCR experiments.

locus id	10'			90'			description	gene name
	p-value	FDR	fold-change	p-value	FDR	fold-change		
PP1387	<0.01	0.06	2.95	<0.01	0.01	16.00	transcriptional regulator TtgR	ttgR
PP1427				<0.01	0.01	5.72	RNA polymerase sigma-H factor AlgU	algU
PP3159	<0.01	0.02	2.78	0.04	0.44	1.85	benABC operon transcriptional activator BenR	benR
PP3168	0.01	0.22	2.23	<0.01	0.02	3.07	benzoate-specific porin	benF
PP3715	<0.01	0.01	4.26	<0.01	0.03	2.95	muconate cycloisomerase	catB

RESULTS AND DISCUSSION

The transcriptome analysis of the stress response due to benzoate pulse after 10 and 90 minutes revealed 95 and 324 significantly differentially expressed genes respectively. Genes were considered to be significantly differentially regulated if p-value ≤ 0.05, FDR ≤ 0.1 and fold-change ≥ 2.0. Genes were selected for RT/PCR experiments (see Table 3.2) due to following characteristics. Firstly, genes were selected because they are well characterized to be involved in the degradation and metabolism of benzoate (*benR*, *benF* and *catB*) and furthermore were already induced 10 minutes after setting the pulse. Secondly, two additional genes (*ttgR* and *algU*) were chosen because of their relative high induction 90 minutes after the pulse. *TtgR*, which that regulates the expression of two efflux pumps, as a repressor of *ttgABC* and an inducer of *ttgDEF*, was previously described to be induced in the presence of toluene (Duque *et al.*, 2001). Hence, as benzoate is an intermediate in the degradation of toluene, *ttgR* could also be involved in the biodegradation of other aromatics. Benzoate as an aromatic acid is a cell wall stressor and thus it is likely that *algU*, recently demonstrated to be involved in cell wall stress (Wood & Ohman, 2009), might also be important in the stress response to high concentrations of benzoate.

The second chemostat was run as described above. Samples for subsequent RNA preparation and RT/PCR examination were taken from steady-state condition and 5, 10, 20, 30, 40, 50, 60, 80, 100, 120, 140, 160, 180 and 210 minutes after the 45 mM benzoate pulse to monitor the stress response over a time period of three hours according to the selected candidate genes. The RT/PCR results can be seen in Figure 3.15. According to the protocol by Bremer *et al.* (1992), the transcript amount according to the RT/PCR results was calculated (see Table 3.3) for all but *catB* transcripts and visualized in Figure 3.16. The RT/PCR result obtained for *catB* transcript showed an inconsistent pattern during the time-course and was therefore excluded from transcript calculation.

RESULTS AND DISCUSSION

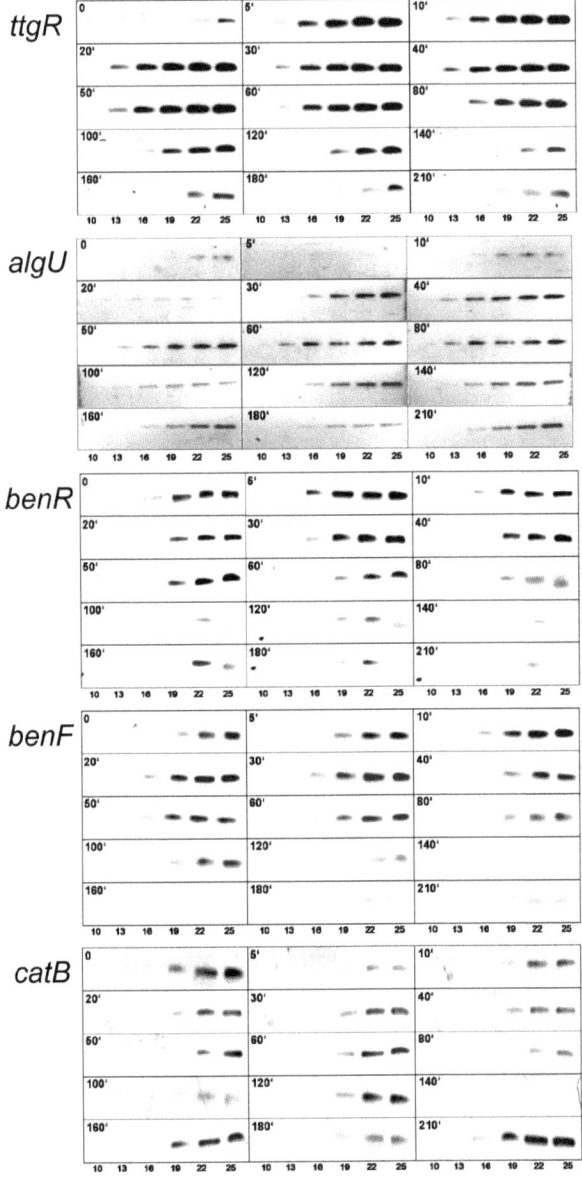

Fig. 3.15 RT/PCR results from first time course experiment for selected candidate genes. The pictures show the first appearance of transcript derived from candidate genes that were selected according to the initial transcriptome data. Time course was monitored from 5 to 210 minutes after benzoate pulse implementation at 14 selected time points. The control sample taken at steady state condition is here marked with 0. Aliquots of PRC reactions were taken firstly after 10 cycles and then in intervals of 3 up to 25 cycles.

Table 3.3 Calculated transcript amounts (Bremer et al., 1992) for the selected candidate genes according to their first appearance in the RT/PCR reactions over a time of 210 minutes.

time	0 *	5'	10'	20'	30'	40'	50'	60'	80'	100'	120'	140'	160'	180'	210'
ttgR															
cycle	20	12	11	10	10	10	10	11	13	13	16	18	18	19	19
mRNA (amol)	5.26	526	936	1660	1660	1660	1660	936	295	295	52.6	16.6	16.6	9.36	9.36
benF															
cycle	16	15	13	13	14	16	16	16	16	18	19	19	19	19	19
mRNA (amol)	52.6	93.6	295	295	166	52.6	52.6	52.6	52.6	16.6	9.36	9.36	9.36	9.36	9.36
algU															
cycle	16	22	19	19	16	13	12	10	13	13	16	16	16	16	16
mRNA (amol)	52.6	1.66	9.36	9.36	52.6	295	526	1660	295	295	52.6	52.6	52.6	52.6	52.6
benR															
cycle	16	14	15	16	16	16	16	16	16	19	19	19	19	19	22
mRNA (amol)	52.6	166	93.6	52.6	52.6	52.6	52.6	52.6	52.6	9.36	9.36	9.36	9.36	9.36	1.66

* The control sample taken at steady state condition is here marked with 0.

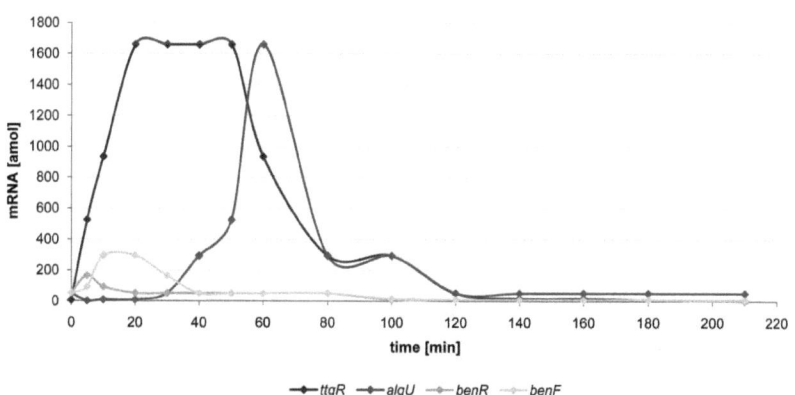

Fig. 3.16 Visualization of detected transcript amounts (amol RNA) due to RT/PCR results for selected genes *ttgR, algU, benR* and *benF* over time course of 210 minutes.

The graphs displaying the calculated transcript amount revealed two phases of stress response. The benzoate specific genes *benR* and *benF* were induced directly after the benzoate pulse, but the transcript amount declined already after 20 minutes. *TtgR* was also induced immediately and reached a plateau of detected transcript from 20 to 60 minutes. A delayed stress response was observed for *algU* where the transcript amount was increasing not before 30 minutes after the pulse. A peak of *algU* transcript was observed after 60 minutes with a rapid decrease thereafter. The transcript level reached a comparable value to the steady-state condition after 40 minutes for the benzoate specific genes *benR* and *benF* and after 120 minutes for *ttgR* and *algU*.

The comparatively low amount of detected transcript for *benR* and *benF* can be ascribed to the experimental setup. Since the growth medium was supplemented with 8.6 mM sodium benzoate as sole carbon source, the cells were adapted to a low concentration of benzoate and hence, genes involved in the ß-ketoadipate pathway for benzoate degradation were already induced. However, the induction of *benR* and *benF* specifically after benzoate pulse is reasonable. Firstly, the enhanced degradation of benzoate via the ß-ketoadipate pathway by *benR* was stimulated since it is an activator for *benABC* converting benzoate to catechol, the first step of degradation (Cowles *et al.*, 2000). Secondly, *benF*, a benzoate specific porin, extrudes benzoate and thus helps to reduce the benzoate concentration within the cell (Nishikawa *et al.*, 2008). Similarly to *benR* and *benF*, the induction of *ttgR* here might lead to the induction of efflux pumps for the extrusion of benzoate, though only *ttgA* (*ttgB* did not fulfil the statistical criteria) was induced 90 minutes after the benzoate pulse according to the initial transcriptome data. *TtgR* was already described in *P. putida* DOT-T1E to regulate *ttgABC* that is induced in the presence of antibiotics (Terán *et al.*, 2003) and *ttgDEF* when exposed to toluene and other aromatic solvents (Duque *et al.*, 2001).

According to the RT/PCR results, eight time points for the time course series of stress response to a high-dosed benzoate pulse were chosen, namely 2, 8, 20, 40, 60, 80, 110 and 180 minutes, to monitor the immediate stress response (2 to 20 minutes), the adaptation phase (20 to 60 minutes) according to the decreasing amount of transcript of benzoate specific genes, and late stress response (60 to 180 minutes).

Though, the chemostat experiment for transcriptional time course monitoring was already run, the results for transcriptome analysis and growth parameter derived from the chemostat are still pending at the time of writing, and thus cannot be included in the analysis.

3.5 Deep RNA Sequencing

The introduction of deep sequencing technologies has changed scientific approaches in the field of genomic research by providing a potential tool to address more global biological questions (Wang et al., 2009). This new platform had already been applied in various studies including identification of genome diversity (Klockgether et al., 2010), transcriptome profiling and identification of small RNAs (Filiatraut et al., 2010; Sharma et al., 2010; Oliver et al., 2009).

The deep sequencing technology from Illumina allows simultaneous sequencing of millions of fragments based on reversible terminator chemistry yielding millions of short reads that can be then mapped to a reference genome, and is therefore also suitable for transcriptome analysis. The advantage hereby is that even low abundant transcripts can be detected and the analysis is not limited to predefined sequences.

The biological goal of this study was the identification of low abundant transcripts with significantly different expression between two experimental conditions. These lowly expressed genes are generally missed with microarrays due to low hybridization intensities and thus signals may be missed in the background noise. Major interest was directed towards the intergenic regions to reveal previously overlooked ORFs and small RNAs. Polycistronic transcripts were used to revise the operon predictions available for *P. putida* KT2440 (www.pseudomonas.com).

The Illumina GA20 Genome Analyzer platform (GATC, Konstanz) was used for deep sequencing of the same samples that were also applied to two different microarray platforms (2.1.2.1 and 2.1.2.2). The results derived from the transcriptome analysis were used for a comparative study of the performance of the three transcriptome platforms.

For this, 5 mL LB broth were inoculated from a frozen stock culture of *P. putida* wild type, and incubated at 30°C for 8 hours at 250 rpm. An aliquot of 0.2 mL was added to 20 mL M9 medium supplemented with 15 mM succinate in a 100 mL flask and incubated overnight at 30°C. For RNA extraction, bacteria were grown in a 1.5 L batch culture (M9 + 15 mM succinate) using the BioFlo 110 Fermenter to assure constant pH, aeration and agitation. When cultures reached mid-exponential phase (OD ~ 0.8) the temperature was decreased from 30°C to 10°C. Three samples, each for parallel RNA extraction, were subsequently taken immediately before temperature downshift (30°C) and two hours after the media was cooled to 10°C.

RESULTS AND DISCUSSION

3.5.1 Evaluation of alignment quality and transcript coverage

The reads provided by GATC that passed the default signal quality filter and were not aligned by ELAND (Illumina) to *P. putida* rRNA genes were used for gene expression analysis.

To generate a summary table comparable to an output file given by the BRB ArrayTool software (Simon *et al.*, 2007; http://linus.nci.nih.gov/BRB-ArrayTools.html) for Progenika and Affymetrix gene expression analysis, the remaining reads were further processed. The reads were aligned on the whole *P. putida* genome sequence (NC_002947) using the Bowtie software (Langmead *et al.*, 2009) for additional quality control and to remove rRNA reads that were not excluded by the previous ELAND software. From the 3' end four nucleotides were trimmed from each read and a seed size of 28 bp was used, in which two mismatches were permitted. The quality mismatch sum was 100 and results were output in SAM format (command line: bowtie -t putida -l 28 -e 100 --best --sam -3 4 -n 2 -p 7).

A summary table was generated with the integrative web analysis tool Galaxy (Giardine *et al.*, 2005). The functions "coverage" and "join" were used respectively to summarize 1) coverage of each ORF in terms of total bp and proportion covered by reads and 2) the number of reads mapped to each ORF. Reads mapped to ORFs had at least 1 bp overlap with the ORF. The two datasets for 30°C and 10°C differed in the absolute number of both total reads and reads that mapped to the genome. In addition, genes differ greatly in length, therefore reads were normalized as follows: The ORF length was standardized to 1000 bp and the number of reads to one million reads per experiment (RPKM, reads per kilobase per million, see Mortazavi *et al.*, 2008).

These analyses were then repeated in the same way for the intergenic regions. Coordinates of intergenic regions were gained using the Galaxy (Giardine *et al.*, 2005) function "complement" on the ORF annotation in Galaxy's "interval" format.

Mapped reads were then visualized with the Integrated Genome Browser (http://igb.bioviz.org/).

Table 3.4   Summary of Illumina deep sequencing data for the two cold shock samples.

	Sample 1 (30°C)	Sample 2 (10°C)	Software
Total number of reads	15699292	17440283	IGA Pipeline 0.2
QC reads	487	524	
Number of reads mapped to ribosomal operons	14280622	15774701	ELAND
Number of non-ribosomal reads	1418183	1665058	
Reads aligned to *P. putida* genome	366990	460289	Bowtie
Thereof not aligned to rRNA	279080	339778	
Reads mapped to ORFs	246999	260915	Galaxy
Reads mapped to IGRs	65858	118310	
Overlapping reads	33777	39447	

3.5.2 Low and high abundant transcripts

Deep sequencing of transcriptome samples is an effective tool for the analysis of differential gene expression since the sequencing depth allows the detection of low abundant transcripts that can be easily missed with microarray techniques. Microarrays show only the relative abundance of transcripts as the presence of a transcript is reflected by signal intensity. Thus, low abundant transcripts might not be detected, as their respective signal intensities are below the detection limit. Abundant transcripts might otherwise be linked to signal saturation and thus differences in their gene expression are hardly detectable.

In this analysis, genes with an RPKM (reads per kilobase per million) of 5000 or higher were considered to be highly abundant. Lowly expressed genes were determined to have a RPKM less than 10. Gene expression was considered to be significant if $RPKM_{30°C} > RPKM_{10°C} + 3\sqrt{RPKM_{10°C}}$ (or vice versa). The 99% confidence interval for the real value N of a Poison-distributed parameter λ is given by $N = N_{emp} \pm 3\sqrt{N_{emp}}$, whereby N_{emp} represents the experimentally determined (empiric) counts.

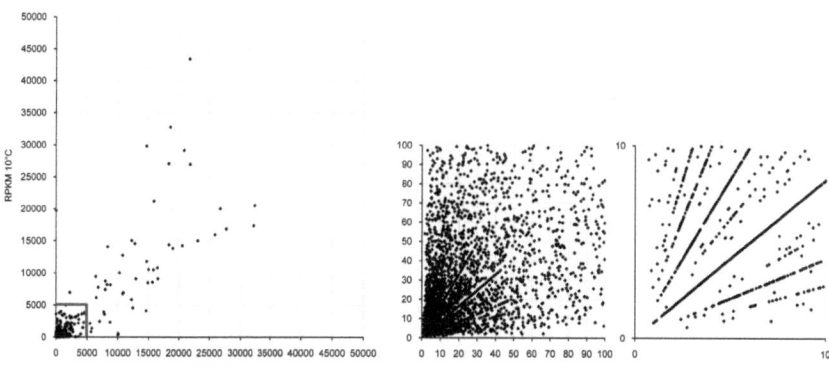

Fig. 3.17 Quantification of expression levels of Illumina cDNA sequencing results. Left diagram shows high abundant transcripts with normalized read numbers higher than 5000 RPKM (reads per kilobase and million). Genes were defined to be significantly expressed if $RPKM_{10°C} > RPKM_{30°C} + 3\sqrt{(RPKM_{30°C})}$ or vice versa. Genes significantly differentially expressed that represent highly abundant transcripts are listed in Table 3.5. The two other pictures display a zoomed in version with low abundant transcripts expressed less than 100 RPKM (middle) and 10 RPKM (right) respectively. Significantly differentially expressed genes that represent low abundance transcripts with a fold-change ≥ 2.0 are listed in Table 3.6.

RESULTS AND DISCUSSION

Table 3.5 Summary of highly abundant transcripts with an RPKM of 5000 or higher in at least one sample.

locus id	RPKM 30°C	RPKM 10°C	fold-change	gene name	description
PP0440	7737.07	3905.22	-1.98	tuf-1	translation elongation factor Tu
PP0444	10863.03	6936.40	-1.57	rplA	ribosomal protein L1
PP0445	18360.32	14421.19	-1.27	rplJ	ribosomal protein L10
PP0446	14656.76	11796.59	-1.24	rplL	ribosomal protein L7/L12
PP0449	10759.26	6782.62	-1.59	rpsL	ribosomal protein S12
PP0450	16543.71	9136.13	-1.81	rpsG	ribosomal protein S7
PP0451	15651.26	8586.67	-1.82	fusA-1	translation elongation factor G
PP0452	12188.29	5849.20	-2.08	tuf-2	translation elongation factor Tu
PP0454	12868.41	9115.68	-1.41	rplC	ribosomal protein L3
PP0455	16469.63	10833.74	-1.52	rplD	ribosomal protein L4
PP0456	26736.18	20070.17	-1.33	rplW	ribosomal protein L23
PP0457	15002.48	10554.44	-1.42	rplB	ribosomal protein L2
PP0458	18997.47	13859.33	-1.37	rpsS	ribosomal protein S19
PP0459	15811.41	10522.47	-1.50	rplV	ribosomal protein L22
PP0460	20544.35	14302.72	-1.44	rpsC	ribosomal protein S3
PP0461	23095.59	15050.43	-1.53	rplP	ribosomal protein L16
PP0462	32415.03	20571.36	-1.58	rpmC	ribosomal protein L29
PP0463	25863.7	15998.96	-1.62	rpsQ	ribosomal protein S17
PP0464	27662.69	16866.84	-1.64	rplN	ribosomal protein L14
PP0465	32214.57	17368.04	-1.85	rplX	ribosomal protein L24
PP0466	14857.98	8507.14	-1.75	rplE	ribosomal protein L5
PP0468	10740.46	12778.46	1.19	rpsH	ribosomal protein S8
PP0469	21808.44	27012.47	1.24	rplF	ribosomal protein L6
PP0470	12213.59	15077.08	1.23	rplR	ribosomal protein L18
PP0471	15887.91	21249.18	1.34	rpsE	ribosomal protein S5
PP0472	18323.18	27106.62	1.48	rpmD	ribosomal protein L30
PP0473	8347.04	14098.40	1.69	rplO	ribosomal protein L15
PP0474	14647.78	29837.86	2.04	secY	Sec-dependent secretion protein SecY
PP0475	21715.43	43461.47	2.00	rpmJ	ribosomal protein L36
PP0476	18590.37	32795.72	1.76	rpsM	ribosomal protein S13
PP0477	20835.99	29173.76	1.40	rpsK	ribosomal protein S11
PP0478	12726.14	14596.83	1.15	rpsD	ribosomal protein S4
PP1464	7971.67	7416.46	-1.07	trmD	tRNA (guanine-N1)-methyltransferase
PP1691	16.00	19708.26	1231.77		conserved hypothetical protein
PP1772	12390	4569.46	-2.71	rpsA	ribosomal protein S1
PP1911	2224.73	6969.65	3.13	rpmF	ribosomal protein L32
PP2466	7979.85	8757.52	1.10	infC	translation initiation factor IF-3
PP2467	8777.98	8181.54	-1.07	rpmI	ribosomal protein L35
PP4185	10032.15	376.21	-26.67	sucD	succinyl-CoA synthetase, alpha subunit
PP4186	10061.23	578.02	-17.41	sucC	succinyl-CoA synthetase, beta subunit
PP4192	5481.91	2135.35	-2.57	sdhD	succinate dehydrogenase, hydrophobic membrane anchor
PP4377	1541.61	5004.98	3.25		flagellin FlaG, putative
PP4709	6400.95	9467.12	1.48	rpsO	ribosomal protein S15
PP4876	6480.92	6167.71	-1.05	rpsR	ribosomal protein S18
PP4877	6820.73	7755.93	1.14	rpsF	ribosomal protein S6
PP5282	7849.64	3554.17	-2.21	rpmB	ribosomal protein L28
PP5414	5645.62	855.55	-6.60	atpG	ATP synthase F1, gamma subunit
PP5415	5780.93	1383.87	-4.18	atpA	ATP synthase F1, alpha subunit
PP5416	8730.71	2333.61	-3.74	atpH	ATP synthase F1, delta subunit
PP5417	14592.01	4126.60	-3.54	atpF	ATP synthase F0, B subunit
PP5418	7026.98	2462.13	-2.85	atpE	ATP synthase F0, C subunit

Table 3.6 Summary of low abundance transcripts with less than 10 RPKM in both samples.

locus id	RPKM 30°C	RPKM 10°C	fold-change	gene name	description
PP0041	1.79	5.90	3.29	cadA-1	cadmium translocating P-type ATPase
PP0144	2.55	8.37	3.28		metalloprotease, insulinase family
PP0145	2.16	7.10	3.29		Na+/Pi cotransporter family protein
PP0222	2.59	8.50	3.28		monooxygenase, DszA family
PP0238	9.36	2.56	-3.65	ssuD	organosulfonate monooxygenase
PP0269	2.15	8.83	4.11		glutamate synthase, large subunit, putative
PP0316	9.72	2.66	-3.65		oxidoreductase, FAD-binding, putative
PP0354	9.25	1.52	-6.09		CBS domain protein
PP0383	2.13	7.00	3.29		tryptophan 2-monooxygenase, putative
PP0613	2.54	8.36	3.29		amidase family protein
PP0619	2.98	9.79	3.29		branched-chain amino acid ABC transporter, periplasmic
PP0763	8.52	1.75	-4.87		medium-chain-fatty-acid CoA ligase
PP0788	1.60	9.21	5.76		conserved hypothetical protein
PP0795	2.06	6.76	3.28	fruA	phosphotransferase system, fructose-specific IIBC
PP0906	1.18	8.70	7.37		multidrug efflux RND transporter, putative
PP1104	9.64	2.64	-3.65		aotO-related protein
PP1228	5.20	1.42	-3.65		methyl-accepting chemotaxis transducer
PP1263	7.03	1.92	-3.65		fusaric acid resistance protein, putative
PP1271	2.33	7.64	3.28		multidrug efflux MFS transporter, putative
PP1445	2.69	8.82	3.28	oprB-2	porin B
PP1450	2.11	8.67	4.11		activation/secretion protein, TPS family, putative
PP1467	1.99	6.52	3.28		sodium/hydrogen exchanger family protein
PP1659	2.67	8.79	3.29		conserved hypothetical protein
PP1703	0.88	7.23	8.21		assimilatory nitrate reductase/sulfite reductase, putative
PP1749	2.05	6.75	3.29		acetyltransferase, GNAT family
PP1820	3.00	9.84	3.28		benzoate transport protein
PP1889	1.44	5.90	4.10	fimD	type 1 pili subunit FimD
PP2038	6.33	1.73	-3.65		long-chain-fatty-acid--CoA ligase, putative
PP2069	2.52	8.28	3.29		multidrug efflux MFS outer membrane protein, putative
PP2185	1.24	7.15	5.76		formate dehydrogenase, alpha subunit, putative
PP2193	7.30	2.40	-3.04		outer membrane ferric siderophore receptor
PP2195	9.88	2.71	-3.65		periplasmic polyamine-binding protein, putative
PP2212	2.42	7.95	3.28		leucine-rich repeat domain protein
PP2287	0.76	6.87	9.05		phage internal core protein
PP2293	2.12	8.72	4.11		DNA maturase B
PP2437	2.88	9.46	3.29		acyl-CoA dehydrogenase, putative
PP2560	6.19	1.70	-3.65		transport protein HasD, putative
PP2566	0.71	9.28	13.08		leucine-rich repeat domain protein
PP2605	2.12	6.97	3.29		conserved hypothetical protein
PP2632	2.22	7.30	3.29		conserved hypothetical protein
PP2641	1.99	6.54	3.29		iron-sulfur cluster-binding protein
PP2645	2.59	7.46	2.88	mgtB	magnesium-translocating P-type ATPase
PP2703	2.23	7.33	3.29		citrate MFS transporter, putative
PP2778	2.81	9.22	3.28		3-oxoacyl-(acyl-carrier-protein) synthase II
PP2800	8.58	2.35	-3.65		diaminobutyrate-2-oxoglutarate transaminase, putative
PP2829	2.64	8.67	3.28		transcriptional regulator, GntR family
PP2839	1.57	9.00	5.74		conserved hypothetical protein
PP2845	6.31	1.73	-3.65	ureC	urease, alpha subunit
PP2853	2.80	9.20	3.28		conserved hypothetical protein
PP2910	2.87	9.42	3.28		conserved hypothetical protein
PP2918	3.47	9.97	2.87		trehalose synthase, putative
PP2944	1.98	9.73	4.92		sensor histidine kinase
PP3184	1.06	9.54	9.00		hypothetical protein
PP3187	8.50	2.33	-3.65	codB	cytosine transporter
PP3325	1.71	9.86	5.76		outer membrane ferric siderophore receptor, putative
PP3330	3.28	9.42	2.87		TonB-dependent siderophore receptor
PP3356	6.08	1.66	-3.65	fcs	feruloyl-CoA-synthetase

Table 3.6 continued

locus id	RPKM 30°C	RPKM 10°C	fold-change	gene name	description
PP3357	7.42	2.03	-3.65	vdh	vanillin dehydrogenase
PP3383	2.01	8.25	4.10		gluconate dehydrogenase, putative
PP3455	9.15	2.50	-3.65		multidrug efflux RND membrane fusion protein
PP3456	1.15	4.70	4.09		multidrug efflux RND transporter
PP3466	6.29	1.72	-3.65		ABC efflux transporter, permease/ATP-binding protein
PP3515	1.72	7.06	4.11	hyuA	hydantoin utilization protein A
PP3529	7.88	2.16	-3.65		monooxygenase, putative
PP3545	2.16	7.09	3.28		sensory box histidine kinase/response regulator
PP3630	2.67	8.77	3.28		porin, putative
PP3718	7.61	2.08	-3.65		aminotransferase, class III
PP3754	9.08	2.49	-3.65		beta-ketothiolase
PP3845	9.69	2.65	-3.65		ABC transporter, periplasmic polyamine-binding protein
PP3881	1.98	9.77	4.93		phage terminase, large subunit, putative
PP4035	2.40	7.90	3.29		transporter, NCS1 nucleoside transporter family
PP4053	5.17	1.06	-4.87		glycosyl hydrolase, putative
PP4083	1.81	7.42	4.10		
PP4155	2.93	9.63	3.29		conserved hypothetical protein
PP4212	8.84	1.81	-4.87		conserved hypothetical protein
PP4221	7.14	0.90	-7.91		non-ribosomal peptide synthetase
PP4232	8.88	2.43	-3.65		cytochrome c family protein
PP4243	4.70	0.91	-5.17		pyoverdine synthetase
PP4278	7.39	2.02	-3.65	xdhA	xanthine dehydrogenase, XdhA subunit
PP4297	6.06	1.66	-3.65	gcl	glyoxylate carboligase
PP4603	2.70	8.86	3.28		ISPpu9, transposase
PP4651	2.50	8.20	3.28	cioA	ubiquinol oxidase subunit I, cyanide insensitive
PP5122	2.24	7.37	3.29		oxidoreductase, GMC family
PP5242	9.23	3.50	-2.64		sensor histidine kinase/GAF domain protein
PP5258	7.21	1.98	-3.65		aldehyde dehydrogenase family protein
PP5361	2.97	9.75	3.28		CobW/P47K family protein

Highly abundant transcripts

In total 51 genes showed high abundant transcripts with 5000 RPKM in at least one sample and were considered to have a significant differentially gene expression under the tested conditions, 16 of which displayed a fold-change ≥ 2.0 (Table 3.5). However, 39 of these genes (76.5 %) encoded for ribosomal proteins or translation elongation factors and were equally expressed under both conditions. This finding is unsurprising, since these genes are the key players in transcription, translation and ribosomal structure and biogenesis. Another subset of genes was significantly down-regulated at 10°C and could be assigned to energy metabolism, either citric acid cycle (*sucCD* and *sdhD*) or oxidative phosphorylation (*atpGAHFE* operon). This is in agreement with knowledge about general cold-shock response of bacteria, where the central metabolism is down-regulated due to a temperature down-shift.

Interestingly, two genes not associated with cold-stress response so far were significantly up-regulated. The gene PP4377, encoding a putative flagellin FlaG protein, is located in the conserved region inbetween *fliC* and *fliD*, which encode the flagella motor and flagella Cap proteins of the flagella assembly complex. The second gene, PP1691, was the most up-regulated gene found. It is worthwhile mentioning that PP1691 was found to be the highest up-regulated gene by all three transcriptome analysis approaches. The three genes are predicted to be organized in an operon and conserved hypothetical proteins with no annotated function, but the data suggest that this gene cluster plays a major role in the response to a rapid temperature down-shift and the survival of the bacterium.

Low abundance transcripts

A very high number of low abundant transcripts indicating little to no expression were detected. In total 1316 genes with coverage of less than 10 RPKM were identified. Among these genes, still 86 displayed a differential expression of at least two-fold. In comparison, Affymetrix microarrays detected only four and the Progenika arrays none of the 86 low abundance genes, using a defined threshold of p-value ≤ 0.05, FDR ≤ 0.05 and fold-change ≥ 2.0. This result shows the advantage of deep RNA sequencing in comparison to microarray platforms, as low abundance transcripts are difficult to detect on a microarray due to resulting low signal intensities that can easily disappear in background noise or fail to pass thresholds in the statistical analysis. Deep RNA sequencing indicates that there is differential expression even among low abundance transcripts that is not detected with microarrays. This might play an important role in adaptation processes to changing environments.

The identification of low abundant transcripts differentially regulated in response to cold shock revealed a high number of transporters. Among the 86 identified low abundant transcripts, 15 code for transporters, two for porins and another three for outer membrane associated proteins, reflecting an alteration of the cell membrane. When cells are exposed to a temperature-downshift the membrane fluidity decreases, and thereby affects membrane-associated functions such as transport and secretion. The high number of transporters among the low abundant transcripts supports the hypothesis that the maintenance of the membrane is one major response to cold shock. These data furthermore indicate that the cell wall maintenance is subject to many fine-tuning processes that are hardly detectable with microarrays.

3.5.3 Finding unpredicted ORFs, small RNAs and unclassified transcripts in intergenic regions

The *Pseudomonas putida* KT2440 chromosome is one of the best studied bacterial genomes. Nevertheless, recent studies in *Pseudomonas* and *Escherichia coli* revealed a high number of small RNAs and also small proteins previously missed with classical transcriptomic or proteomic studies (Livny *et al.*, 2006; Sonnleitner *et al.*, 2008; Hemm *et al.*, 2010; Filiatrault *et al.*, 2010; Hobbs *et al.*, 2010), demonstrating that there is a wealth of information in intergenic regions to be discovered and indicating weaknesses in the existing genome annotations.

In this study, intergenic regions were investigated for their expression to identify previously undiscovered transcripts that could potentially be assigned to novel protein coding open reading frames. Furthermore, presence and expression of small RNAs already described in other *Pseudomonas* species (Livny *et al.*, 2006; González *et al.*, 2008; Sonnleitner *et al.*, 2008, 2009) were identified by sequence similarity and analyzed for transcript expression.

Sequences of intergenic regions with considerable number of transcripts were analysed with gene prediction software to identify possible protein coding open reading frames. GeneMark (http://exon.biology.gatech.edu/) predictions were performed using GeneMark.hmm 2.4 and the GeneMark.hmm heuristic version, where gene prediction is performed with a pre-computed prediction model based on the reference genome (Lukashin & Borodovsky, 1998) or with a heuristic approach that can be used for gene predictions in novel genomes (Besemer & Borodovsky, 1999), respectively. Glimmer (http://www.cbcb.umd.edu/software/glimmer/), another gene prediction model, uses interpolated Markov models for the identification of protein coding regions (Salzberg *et al.*, 1998; Delcher *et al.*, 1999). Where genes were predicted, the sequence was searched using Blastx for homologous genes in the NCBI protein database for Bacteria and Archaea (http://blast.ncbi.nlm.nih.gov/Blast.cgi). Predicted ORFs can vary in length due to the presence of multiple possible start sites. Thus, determination of a putative ORF was based on detecting transcripts which best covered the predicted open reading frame.

In the case of small RNA identification, since they are harder to detect due to their small size and lack of prediction tools due to scarce knowledge of sequence motifs, the approach was carried out inversely to the prediction of protein coding open reading frames. Sequences of previously verified sRNAs in *Pseudomonas* species were used to identify transcripts on the basis of sequence similarity. Since sequence conservation of sRNAs among species is rather low, this approach was limited to sRNAs predicted so far in *Pseudomonas*, and not extended to sRNA sequences in *E.coli* where more than 80 sRNAs have already been detected (Hobbs *et al.*, 2010).

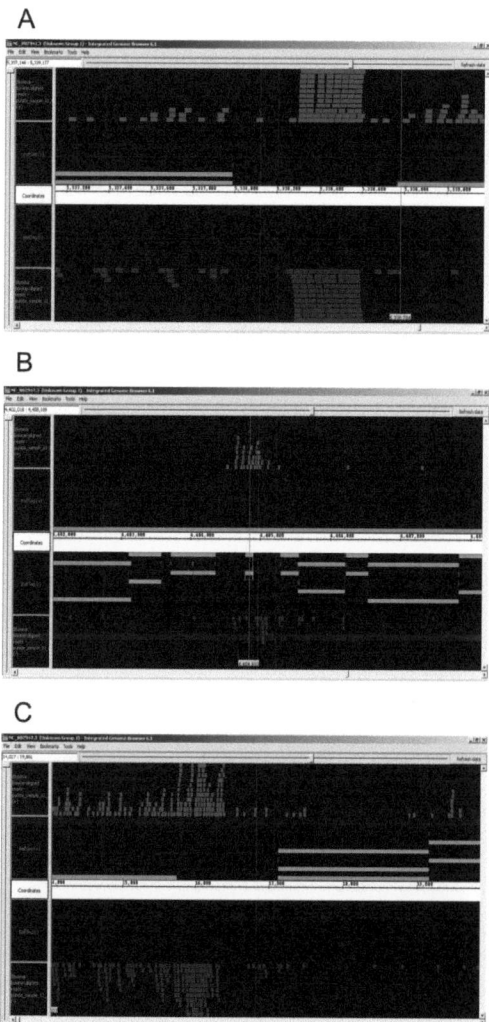

Fig. 3.18 IGB (Integrated Genome Browser) interface showing transcript of A) the sRNA CrcZ at 30°C, B) a previously misannotated gene PP4885 at 30°C and C) a newly identified gene PP0013.1 at 10°C.

RESULTS AND DISCUSSION

Table 3.7 Summary of detected new protein coding open reading frames based upon ORFs with detected transcripts.

#	locus id	start	stop	function	e-value	RPKM 30°C	RPKM 10 °C	fold-change
1	PP0013.1	16589	17227	conserved hypothetical	2.00E-115	31.09	610.75	19.64
2	PP0017.1	22973	23200	hypothetical		77.65	455.59	5.87
3	PP0048.1	56025	55444	conserved hypothetical	9.00E-86	113.75	111.40	-1.02
4	PP0049.1	58089	57640	conserved hypothetical	1.00E-69	21.81	30.71	1.41
5	PP0167.1	193795	193580	putative outer membrane protein, TolC family-like	6.00E-06	76.94	403.71	5.24
	PP0167.2	194084	193881	hypothetical	3.00E-04			
6	PP0168.1	220688	221029	conserved hypothetical	2.00E-11	135.81	557.73	4.11
7	PP0201.1	251820	252062	conserved hypothetical	3.00E-27	21.10	34.66	1.64
	PP0201.2	252132	252770	conserved hypothetical	1.00E-112			
8	PP0305.1	367312	366695	conserved hypothetical	5.00E-58	5.16	59.29	11.50
9	PP0636.1	744476	744916	hypothetical		5.68	349.81	61.10
10	PP0638.1	747946	748764	hypothetical	4.00E-65	37.66	119.52	3.17
11	PP0640.1	751604	751005	hypothetical	8.80E-02	17.99	125.63	6.98
	PP0640.2	751985	752284	hypothetical				
12	PP0651.1	759472	759867	conserved hypothetical	2.00E-06	12.73	57.50	4.52
	PP0651.2	759860	760087	conserved hypothetical	8.00E-16			
13	PP0655.1	763276	763635	conserved hypothetical	1.00E-34	77.15	49.29	-1.57
14	PP0877.1	1017859	1017380	putative transcriptional regulator, LysR family	5.00E-09	34.13	203.21	5.95
15	PP1115.1	1275673	1275113	hypothetical, homologue to PP2294	8.00E-41	174.01	308.72	1.77
16	PP1116.1	1278330	1277962	conserved hypothetical	1.00E-19	7.86	430.59	54.76
17	PP1174.1	1349722	1350087	conserved hypothetical	7.00E-09	5.25	73.36	13.96
18	PP1781.1	1995218	1994979	conserved hypothetical	4.00E-06	49.96	225.69	4.52
	PP1781.2	1995616	1995365	conserved hypothetical	3.00E-13			
19	PP1808.1	2034392	2034673	putative lipoprotein	2.00E-30	46.75	233.14	4.99
20	PP1809.1	2036041	2035622	hypothetical		16.47	90.21	5.48
21	PP1867.1	2090657	2090469	hypothetical	5.00E-03	142.51	777.58	5.46
22	PP1919.1	2163137	2164150	hypothetical		126.00	506.55	4.02
23	PP1920.1	2166153	2165602	hypothetical	2.00E-35	23.03	117.72	5.11
24	PP1935.1	2182187	2182579	conserved hypothetical	6.00E-24	370.96	202.01	-1.84
	PP1935.2	2182593	2183252	conserved hypothetical	1.00E-96			
	PP1935.3	2183212	2183766	conserved hypothetical	3.00E-52			
	PP1935.4	2183940	2184506	putative resolvase	2.00E-19			
25	PP1936.1	2188522	2187608	putative SEC-C motif domain protein	8.00E-33	224.65	232.50	1.03
	PP1936.2	2188752	2188522	hypothetical	2.00E-10			
26	PP1957.1	2216100	2216393	conserved hypothetical	3.00E-05	43.11	305.07	7.08
	PP1957.2	2216487	2217281	conserved hypothetical	5.00E-20			
27	PP1958.1	2218192	2218407	hypothetical	4.00E-03			
	PP1958.2	2218872	2218411	hypothetical		177.48	732.82	4.13
28	PP2026.1	2303141	2302944	hypothetical	5.10E-02	8.16	53.63	6.57
29	PP2062.1	2346563	2345610	hypothetical protein	6.00E-05	17.93	137.46	7.67
30	PP2095.1	2388619	2388404	ribosome modulation factor	9.00E-26	57.20	171.15	2.99
31	PP2127.1	2427820	2427212	putative CheA signal transduction histidine kinase	3.00E-32	3.03	102.00	33.68
32	PP2294.1	2622805	2623161	conserved hypothetical	1.00E-19	58.25	123.73	2.12
33	PP2437.1	2785234	2785593	conserved hypothetical	6.00E-37	4.41	50.74	11.50
34	PP2472.1	2818042	2817200	hypothetical		27.11	37.57	1.39
35	PP2504.1	2851568	2852086	conserved hypothetical	7.00E-54	20.00	19.16	-1.04
36	PP2509.1	2857242	2857700	putative histidine triad (HIT) protein	1.00E-41	514.05	337.06	-1.53
37	PP2541.1	2886990	2886109	YD repeat-containing protein	2.00E-19	23.61	269.32	11.41
38	PP2680.1	3070537	3070731	putative coenzyme PQQ biosynthesis protein A	2.00E-21	263.17	108.08	-2.43
39	PP2984.1	3382497	3382664	hypothetical		10.25	129.06	12.59
	PP2984.2	3382885	3383190	conserved hypothetical	1.00E-31			
	PP2984.3	3383191	3383493	conserved hypothetical	4.00E-30			
40	PP3024.1	3411246	3411470	hypothetical	1.00E-05	324.49	635.12	1.96
	PP3024.2	3411424	3412008	hypothetical	8.00E-21			
41	PP3066.1	3447310	3447792	putative phage protein	1.00E-44	22.90	67.70	2.96
	PP3066.2	3448242	3448565	hypothetical	1.00E-01			
42	PP3067.1	3450583	3450218	hypothetical	1.30E-02	61.34	134.35	2.19
43	PP3080.1	3465989	3465840	hypothetical	3.00E-06	58.11	21.21	-2.74
44	PP3090.1	3482120	3481212	conserved hypothetical	4.00E-171	7.95	68.25	8.62

Table 3.7 continued

#	locus id	start	stop	function	e-value	RPKM 30°C	RPKM 10 °C	fold-change
45	PP3100.1	3497593	3496829	conserved hypothetical	1.00E-08	22.82	117.83	5.16
	PP3100.2	3498850	3498299	conserved hypothetical	2.00E-95			
46	PP3101.1	3501611	3502309	hypothetical, homologue to PP3105	2.00E-45	7.98	180.83	22.67
	PP3101.2	3502340	3502546	hypothetical, homologue to PP3106	3.00E-09			
	PP3101.3	3503012	3502701	hypothetical				
	PP3101.4	3503302	3503021	hypothetical				
47	PP3108.1	3515481	3515957	hypothetical	3.00E-32	255.94	185.63	-1.38
	PP3108.2	3516158	3516916	hypothetical	2.20E-02			
	PP3108.3	3517092	3517622	YD repeat-containing protein	1.00E-126			
48	PP3109.1	3518850	3518371	conserved hypothetical	7.00E-35			
	PP3109.2	3519593	3520192	hypothetical	9.00E-05	389.92	268.01	-1.45
	PP3109.3	3520258	3520725	putative cytoplasmic protein	1.00E-10			
48	PP3109.4	3521239	3521844	hypothetical	4.00E-20			
49	PP3125.1	3537125	3537349	conserved hypothetical	4.00E-15	20.71	106.33	5.13
50	PP3158.1	3578497	3579228	conserved hypothetical	7.00E-19	9.29	129.70	13.96
51	PP3328.1	3765667	3766539	putative transcriptional regulator, LysR family	3.00E-112	21.50	26.49	1.23
52	PP3547.1	4022609	4022259	hypothetical	2.00E-13	4.15	23.87	5.75
53	PP3603.1	4095647	4096315	conserved hypothetical	2.00E-68	3.88	66.96	17.25
54	PP3682.1	4186128	4186415	putative transposase	6.00E-03	6.29	134.25	21.36
55	PP3684.1	4188656	4189552	putative transcriptional regulator	2.00E-33	52.69	95.22	1.81
56	PP3685.1	4190445	4190053	hypothetical	5.00E-13	7.82	151.01	19.30
57	PP3687.1	4193644	4193955	hypothetical	1.00E-12	19.63	73.28	3.73
	PP3687.2	4194548	4194144	hypothetical	3.00E-26			
58	PP3688.1	4197678	4197145	hypothetical	1.00E-56	41.68	48.40	1.16
	PP3689.2	4195708	4196154	putative DANN-binding protein	1.00E-10			
59	PP3689.1	4199217	4199696	putative ATP-binding protein	6.00E-08	238.00	792.79	3.33
60	PP3692.1	4211072	4210788	hypothetical		284.52	985.90	3.47
61	PP3697.1	4217790	4217035	conserved hypothetical	2.00E-09	18.98	62.37	3.29
62	PP3705.1	4226893	4226213	conserved hypothetical	3.00E-78	295.94	149.59	-1.98
63	PP3706.1	4228360	4227854	conserved hypothetical	6.00E-30	185.01	93.04	-1.99
64	PP3710.1	4232885	4233319	putativetranscriptional regulator, Fis family	3.00E-23	4.71	54.22	11.50
65	PP3791.1	4321135	4321749	ThiJ/PfpI	1.00E-45	39.38	37.73	-1.04
66	PP3898.1	4413328	4414320	hypothetical protein	4.00E-10	882.57	1033.81	1.17
	PP3898.2	4414323	4414889	putative phage membrane protein	6.00E-13			
67	PP3916.1	4424940	4424669	conserved hypothetical	2.00E-14	8.52	76.96	9.03
68	PP3987.1	4497948	4496701	putative galactose mutarotase-like protein	2.00E-14	488.74	931.42	1.91
69	PP4085.1	4618243	4618728	conserved hypothetical	1.00E-42	91.15	514.69	5.65
	PP4085.2	4619304	4619483	conserved hypothetical	7.00E-10			
70	PP4089.1	4624413	4624874	hypothetical		159.76	224.95	1.41
71	PP4093.1	4628763	4629233	hypothetical protein	9.00E-48	43.90	55.48	1.26
	PP4093.2	4629578	4629724	RHS family like protein	3.00E-17			
72	PP4094.1	4630222	4631078	conserved hypothetical	4.00E-17	37.67	230.36	6.11
73	PP4238.1	4812977	4813636	conserved hypothetical	1.00E-04	8.08	19.91	2.46
74	PP4239.1	4813945	4814427	conserved hypothetical	1.00E-04	29.98	67.73	2.26
75	PP4240.1	4815694	4816374	conserved hypothetical	2.00E-92	3.06	77.85	25.46
76	PP4270.1	4856260	4856069	sensory box protein/GGDEF family protein	1.00E-07	12.41	50.96	4.11
77	PP4346.1	4938845	4939294	conserved hypothetical	8.00E-48	7.53	117.48	15.61
78	PP4350.1	4943725	4944639	D-alanine--D-alanine ligase	2.00E-81	13.70	45.02	3.29
79	PP4410.1	5007017	5007616	hypothetical		13.70	45.02	3.29
80	PP4415.1	5012406	5012005	hypothetical	2.00E-14	10.67	46.02	4.31
81	PP4448.1	5045301	5044960	hypothetical	9.00E-28	5.23	107.41	20.53
82	PP4450.1	5046781	5047017	hypothetical	2.00E-18	17.87	165.14	9.24
	PP4450.2	5047468	5047788	hypothetical	2.00E-07			
83	PP4451.1	5050409	5049780	conserved hypothetical	5.00E-03	24.16	163.03	6.75
	PP4451.2	5051058	5050414	hypothetical, MFS like	5.00E-22			
84	PP4467.1	5073613	5073230	conserved hypothetical	9.00E-41	15.86	110.73	6.98
	PP4467.2	5074119	5073673	hypothetical	2.00E-20			
	PP4467.3	5074561	5074169	conserved hypothetical	3.00E-41			

Table 3.7   continued

#	locus id	start	stop	function	e-value	RPKM 30°C	RPKM 10 °C	fold-change
85	PP4491.1	5103175	5103384	conserved hypothetical	7.00E-18	15.20	474.56	31.21
86	PP4533.1	5148537	5148268	conserved hypothetical	3.00E-25	106.35	221.74	2.08
87	PP4535.1	5151921	5151460	hypothetical	7.00E-20	44.02	69.90	1.59
	PP4535.2	5152241	5151921	hypothetical	8.70E-02			
88	PP4546.1	5168080	5168295	conserved hypothetical	4.00E-11	7.11	70.07	9.86
89	PP4613.1	5237168	5237515	conserved hypothetical	4.00E-21	26.06	26.76	1.03
90	PP4629.1	5253118	5253735	conserved hypothetical	3.00E-38	75.78	195.61	2.58
91	PP4633.1	5256505	5256846	putative HopJ type III effector protein	2.00E-20	37.82	186.37	4.93
92	PP4714.1	5361597	5361199	preprotein translocase subunit SecG	8.00E-37	1441.71	3240.87	2.25
93	PP4739.1	5391285	5391046	conserved hypothetical	1.00E-27	62.99	902.47	14.33
	PP4739.2	5391893	5391624	hypothetical	1.00E-03			
94	PP4742.1	5398728	5399156	conserved hypothetical	3.00E-40	328.52	545.95	1.66
95	PP4947.1	5635545	5634802	conserved hypothetical	1.00E-89	23.42	19.24	10.68
96	PP5129.1	5852610	5852284	conserved hypothetical	2.00E-14	9.24	166.88	18.07
97	PP5201.1	5934113	5934331	conserved hypothetical	8.00E-22	49.63	65.22	1.31
98	PP5208.1	5941880	5941446	conserved hypothetical	2.00E-59	91.99	154.90	1.68
99	PP5238.1	5972772	5972990	hypothetical, homologue to PP4885	3.00E-08	22.54	86.66	3.85
	PP5238.2	5972954	5973679	conserved hypothetical	7.00E-42			
	PP5238.3	5973676	5974032	conserved hypothetical	2.00E-37			
100	PP5266.1	6016693	6015833	hypothetical	1.00E-28	8.86	422.18	47.64
101	PP5379.1	6131818	6131501	conserved hypothetical	2.00E-07	319.05	131.03	-2.43
102	PP5384.1	6137096	6137446	conserved hypothetical	4.00E-22	7.51	67.87	9.03
103	PP5388.1	6144454	6143963	hypothetical	7.00E-12	24.61	106.86	4.34
104	PP5395.1	6151818	6152393	conserved hypothetical	3.00E-63	444.65	378.55	-1.17
	PP5395.2	6152410	6152898	hypothetical	4.00E-29			
105	PP5401.1	6158375	6158653	hypothetical		54.33	340.58	6.27

Finding new protein coding open reading frames

The gene prediction analysis revealed 105 intergenic regions to encode for 143 hypothetical proteins, of which 14 did not show homology to any protein in the database. The other predicted protein coding sequences disclosed similarity to yet annotated proteins in the database with an e-value below $1e^{-10}$. Proteins with lower confidence values displayed either constantly high transcript levels under both tested conditions or a significant expression change of at least four-fold. Newly predicted hypothetical proteins with their predicted start and stop position, transcript coverage and e-value are listed in Table 3.7.

Identification of sRNA genes

To date, out of about 70 candidate sequences, over 40 different sRNA genes have been verified by Northern blot in *Pseudomonas* species; about 40 in *P. aeruginosa* (Livny et al., 2006; Gonzaléz et al., 2008; Sonnleitner et al., 2008), among others the *crcZ* gene in *P. fluorescens* (Sonnleitner et al., 2009) and another five, among others, in *P. syringae* (Filiatraut et al., 2010). For the analysis of conserved sequences in intergenic regions to reveal sRNA genes in *Pseudomonas putida*, all candidate sequences described in these studies were screened. About half of the yet verified sRNA

RESULTS AND DISCUSSION

genes could be detected in *Pseudomonas* putida based on sequence similarity and transcript levels. The 22 sRNA genes with their genomic localization are listed in Table 3.8.
All but three sRNA genes show only low to moderate expression differences due to cold-shock. The *crcZ* gene and the TPP riboswitch were significantly down-regulated, while the P1 sRNA was significantly up-regulated, suggesting a central role in cold-shock response.

Unclassified transcripts
Furthermore, apart from novel predicted hypothetical proteins and identified sRNAs, a high number of unclassified transcripts were detected that could not be assigned either to protein coding open reading frames or sRNA ORFs. Further bioinformatic analysis or experiments that were not applied in this study to date would be necessary to determine if transcripts encode proteins or sRNAs, either by identification of secondary structure and predicted termination regions or by analyzing transcripts for protein or RNA activity. Unclassified transcripts which may be assigned to either protein coding regions or sRNA are listed in Table 3.9.

Genes with a predicted open reading frame differing from the observed transcript
Apart from the analysis of expression in intergenic regions, annotated ORFs from GenBank annotation (NC_002947) were screened for possible misannotations. In case of sufficient coverage, but discrepancy between the identified transcript and predicted open reading frame, the region was screened for ORFs with another start codon or placed in another reading frame more suitable to the detected transcript. In this study, 10 genes were found to be potentially misannotated due to the observed transcript coverage. The genes are listed in Table 3.10.

Verification of polycistronic transcripts
Since operons are laborious to verify experimentally, predictions are mainly based on *in silico* analysis and functional relation. Deep RNA sequencing provides a promising tool to identify operons due to transcript coverage of consecutive ORFs and respective intergenic regions. The analysis of polycistronic transcripts derived from 30°C and 10°C samples revealed 196 operons, of which 57 differed from the current prediction. A summary of the 66 identified polycistronic transcripts differing from the prediction are listed in Table 3.11. The list with polycistronic transcripts which verify the prediction can be found in the appendix.

Table 3.8 Summary of identified sRNA genes.

	sRNA	location	RPKM 30°C	RPKM 10°C	fold-change	RFAM prediction[a] start	RFAM prediction[a] stop	length	strand	identified sequence[b] start	identified sequence[b] stop	condition
[4]	Spot42-like (spf)	IGR 0124	106.24	158.65	1.49			127		130354	130540	30°C
2	RsmY	IGR 0371	24.21	19.89	1.17	450801	450927		+	450782	450934	30°C
1	P26	IGR 0447	3891.26	3728.81	-1.04	537405	537470	66	+	537397	537620	30°C
	FMN riboswitch	IGR 0531	71.95	47.28	-1.52	616380	616513	134	+	616476	616551	10°C
4	YybP-YkoY	IGR 0761	66.60	43.76	-1.52	875976	876117	141	+	875987	876059	30°C
	*T44	IGR 1591	340.76	178.11	-1.91	1785128	1785222	94		1785103	1785294	30°C
2	RsmZ	IGR 1625	64.33	137.38	2.14					1822044	1822205	10°C
1	RgsA / P16	IGR 1968	119.44	158.47	1.33	2229914	2229710	195	−	2229891	2229795	30°C
1	P15	IGR 3081	58.11	21.21	-2.74	3466159	3466041	119	−	3466018	3465882	30°C
4	Psr2	IGR3541	1889.08	121.48	-15.56					4013231	4013564	30°C
1	P6	IGR 4487	53.48	76.87	1.44					5096922	5097007	10°C
3	PrrF	IGR 4686	98.98	97.56	-1.01	5325379	5325527	149	+	5325358	5325484	30°C
3	CrcZ	IGR 4697	3110.03	399.96	-7.78				+	5338247	5338612	30°C
4	*S15	IGR4709	1199.16	668.35	-1.79	5354527	5354623	96		5354527	5354623	30°C
1	P31	IGR 4725	186.28	72.00	-2.59					5373151	5373213	10°C
1	P32	IGR 4725	186.28	72.00	-2.59					5373289	5373325	30°C
2	SsrA tmRNA	IGR 4739	7274.98	7220.54	-1.01	5390008	5390370	362	+	5390008	5390370	30°C
	TPP riboswitch	IGR 4923	67.93	6.97	-9.74	5596334	5596228	107	−	5596326	5596192	30°C
1	P1	IGR 5184	37.39	174.03	4.65	5912448	5912629	182	+	5912456	5912679	10°C
	6S RNA	IGR 5203	99.84	236.91	2.37	5934663	5934842	180	+	5934873	5934984	10°C
1	*RnpB / P28	PP1327	580.14	946.00	1.63	1512692	1513067	376	+	1512687	1512986	30°C
4	*Cobalamin	PP1671	21.46	70.49	3.28	1866934	1867155	221		1867038	1867183	10°C

* Transcripts of identified sRNA overlaps with transcripts of up-stream ORF.
# Identified sRNA gene is located in a predicted protein coding ORF.
[1] Livny et al., 2006; [2] Sonnleitner et al., 2008; [3] Gonzaléz et al., 2008; [4] Filiatraut et al., 2010.
[a] Coordinates of the predicted sRNAs in the P. putida KT2440 genome are taken from RFAM (http://rfam.sanger.ac.uk/).
[b] Coordinates of the predicted sRNAs in the P. putida KT2440 genome are according to identified transcripts based on the Illumina cDNA sequencing results.

RESULTS AND DISCUSSION

Table 3.9 Summary of unclassified transcripts encoding either proteins or sRNA genes.

# IGR	IGR start	IGR stop	external 5' 3' boundaries of mapped cDNA reads 30°C		10°C		RPKM 30°C	10 °C	fold-change
*IGR 0014	15750	17133	15710	16483	15710	16407	31.09	610.75	19.64
*IGR 0018	22530	23176	22630	22962	22521	22962	77.65	455.59	5.87
*IGR 0168	193516	194494	194255	194489	194076	194495	76.94	403.25	5.24
IGR 0275	333728	334148	333943	333976	333788	334139	8.53	91.1	10.68
IGR 0314	376722	377324	376790	377321	376739	377162	11.90	68.44	5.75
IGR 0375	454455	454815	454530	454795	454455	454805	209.02	212.56	1.02
IGR 0528	614251	614357	614154	614265	614183	614271	202.82	111.06	-1.83
IGR 0601	707346	707597	707302	707507	707293	707527	656.68	703.53	1.07
IGR 0642	752759	753914			753254	753895	9.31	63.7	6.84
IGR 0752	869585	870175	869586	870072	869603	870153	206.49	154.64	-1.34
IGR 0886	1028737	1029118	1028962	1029071	1028769	1029120	18.81	602.52	32.03
*IGR 1116	1274083	1276657	1274170	1275057	1274140	1274953	174.01	308.72	1.77
			1275983	1276579	1275996	1276635			
*IGR 1117	1277367	1278734			1277373	1277921	7.86	430.59	54.76
			1278576	1278694	1278331	1278722			
IGR 1782	1994745	1996036	1995680	1996036	1995640	1996002	49.96	225.69	4.52
*IGR 1809	2034181	2035254	2034247	2034323	2034247	2034307	46.75	233.14	4.99
			2034715	2034985	2034711	2035239			
*IGR 1920	2162693	2164314	2162740	2162916	2162675	2162924	126.00	506.55	4.02
IGR 1922	2168097	2168785	2168323	2168544	2168207	2168758	15.62	119.78	7.67
*IGR 1936	2182032	2184669	2183828	2183864	2183844	2183907	370.96	202.01	-1.84
*IGR 2063	2345344	2346543	2345494	2345547	2345416	2345612	17.93	137.46	7.67
*IGR 2096	2388260	2389137	2388781	2389104	2388731	2389127	57.20	171.15	2.99
*IGR 2128	2426890	2428073			2426959	2427200	3.03	102	33.68
IGR 2165	2470162	2470433			2470312	2470419	13.22	97.74	7.39
*IGR 2295	2621623	2623407	2621638	2621903	2621633	2622547	58.25	123.73	2.12
IGR 2303	2632929	2633082	2632921	2633100	2632929	2633072	1405.18	596.31	-2.36
*IGR 2438	2784789	2785601			2784791	2784937	4.41	50.74	11.50
IGR_t57	2816886	2817001	2816873	2816991	2816877	2816985	436.22	511.84	1.17
IGR 2477	2823456	2823921	2823463	2823919	2823571	2823606	108.35	19.07	-5.68
*IGR 2510	2856728	2858387	2856843	2857198	2856837	2857186	514.05	337.06	-1.53
			2857758	2858315	2857768	2858344			
*IGR 2542	2885875	2887241	2886872	2887230	2886999	2887229	23.61	269.32	11.41
IGR 2876	3276692	3276975	3276712	3276943			88.63	10.40	-8.52
IGR 3040	3424409	3424921			3424425	3424820	7.00	224.18	32.03
IGR 3067	3447235	3448800	3447901	3448069	3447805	3448102	22.90	67.7	2.96
*IGR 3102	3499780	3504272	3499918	3500004	3499824	3500909	7.98	180.83	22.67
					3501340	3501603			
			3503820	3504226	3503391	3504241			
*IGR 3110	3518312	3521804	3518906	3519455	3518965	3519455	389.92	268.01	-1.45
			3520815	3521238	3520823	3521221			
*IGR 3159	3578320	3579863			3579246	3579526	9.29	129.7	13.96
IGR 3586	4073458	4074082	4073515	4074020	4073623	4074030	57.42	56.6	-1.01
*IGR 3683	4186080	4186650			4186517	4186609	6.29	134.25	21.36
*IGR 3686	4189942	4191774			4190441	4190843	7.82	151.01	19.30
*IGR 3688	4192963	4194971	4193006	4193538	4193025	4193591	19.63	73.28	3.73
IGR 3704	4223884	4224705	4224022	4224664	4223916	4224705	48.01	555.64	11.57
*IGR 3707	4227839	4228788	4228425	4228539	4228368	4228792	185.01	93.04	-1.99
*IGR 3899	4413161	4415057	4413159	4413304	4413159	4413331	882.57	1033.81	1.17
*IGR 3917	4424478	4425740	4424983	4425302	4425079	4425539	8.52	76.96	9.03
IGR 4086	4618219	4619477	4618887	4619231	4618745	4619242	91.15	514.69	5.65
IGR 4095	4630222	4631078	4630288	4631016	4630234	4631028	37.67	230.36	6.11
IGR 4116	4653029	4653520	4653125	4653525	4653135	4653168	211.64	17.98	-11.77
IGR 4245	4832024	4832176	4832022	4832154			306.46	38.78	-7.91
*IGR 4451	5046667	5048271			5047062	5047421	17.87	165.14	9.24
			5047900	5048234	5047861	5048229			
*IGR 4452	5049356	5051432			5049356	5049754	24.16	163.03	6.75
*IGR 4468	5072032	5074743	5072484	5072750	5072072	5073105	15.86	110.73	6.98
*IGR 4492	5103066	5103773	5103432	5103465	5103393	5103540	15.20	474.56	31.21
					5103603	5103766			
IGR 4535	5148875	5149460	5148958	5149458	5148919	5149404	61.25	100.62	1.64
*IGR 4740	5391009	5393057	5391481	5391584	5391300	5391600	62.99	902.47	14.33
			5391930	5392702	5391891	5392784			
IGR 5077	5797890	5798264	5797893	5798107	5797911	5798082	364.07	78.69	-4.63
*IGR 5267	6015689	6016902			6016689	6016901	8.86	422.18	47.64
IGR 5349	6097839	6098030			6097833	6098039	18.76	215.72	11.50

* IGRs which were also found to contain novel ORFs

Table 3.10 Summary of misannotated genes with new predicted open reading frame based upon deep sequencing evidence.

ID	old annotation				new annotation			
	start	stop	length	strand	start	stop	length	encoded protein
PP0373	453510	453668	159	+	453640	454275	636	hypothetical
PP0731	849197	848580	618	-	848580	850415	1836	phosphatidylserine synthase
PP1327	1513067	1512647	421	-	1512692	1513067	376	sRNA gene *rnpB*
PP2063	2346543	2346647	105	+	2345610	2346563	954	hypothetical
PP2146	2449762	2447762	2001	-	2447762	2451079	3318	Snf2/Rad54 family helicase
PP3885	4404899	4404792	108	-	4404520	4405149	630	hypothetical
PP4534	5148875	5148633	243	- / +	5148497	5149039	543	hypothetical
PP5026	5728628	5726949	1680	-	5726949	5728715	1767	glucan biosynthesis protein G
PP5211	5944225	5943668	558	-	5943668	5944318	651	ChaC family protein
PP5402	6160309	6159521	789	-	6157627	6158643	1017	Sir2 family transcriptional regulator

Table 3.11 Summary of operons verified by detected polycistronic transcripts that differ from the current prediction at www.pseudomonas.com[#].

verified operons	gene name	function	functional category	current annotation[#]
PP0058		phospholipid/glycerol acyltransferase		PP0059-0016
PP0059		D,D-heptose 1,7-bisphosphate phosphatase		
PP0060	glyS	glycyl-tRNA synthetase subunit beta		
PP0061	glyQ	glycyl-tRNA synthetase subunit alpha		
PP0064		TPR domain-containing protein	Translation, ribosomal structure and biogenesis	PP0064-0067
PP0065	trkA	potassium transporter peripheral membrane component		
PP0066	sun	sun protein		
PP0067	fmt	methionyl-tRNA formyltransferase		
PP0068	def	peptide deformylase		
PP0112		metal ABC transporter periplasmic-binding protein	ABC transporter system	PP0113-0114
PP0113		ABC transporter, permease protein		
PP0114		ABC transporter ATP-binding protein		
PP0160		TonB-dependent siderophore receptor		PP0161-0163
PP0161		anti-FecI sigma factor, FecR		
PP0162		ECF subfamily RNA polymerase sigma factor		
PP0163		GntR family transcriptional regulator		
PP0196		ABC transporter ATP-binding protein	ABC transporter system	PP0196-0197
PP0197		hypothetical protein		
PP0198		amino acid transporter LysE		
PP0282		amino acid ABC transporter, periplasmic amino acid-binding protein	ABC transporter system	PP0280-0283
PP0283		amino acid ABC transporter ATP-binding protein		
PP0338	aceF	dihydrolipoamide acetyltransferase	Energy production and conversion	no prediction
PP0339	aceE	pyruvate dehydrogenase subunit E1		
PP0341	waaF	lipopolysaccharide heptosyltransferase II	Cell envelope biogenesis, outer membrane lipopolysaccharide metabolism	PP0341-0345
PP0342	waaC	lipopolysaccharide heptosyltransferase I		PP0346-0357
PP0343	waaG	lipopolysaccharide core biosynthesis protein WaaG		
PP0344	waaP	lipopolysaccharide kinase		
PP0345		lipopolysaccharide kinase		
PP0346		lipopolysaccharide kinase		
PP0347		hypothetical		
PP0375		prolyl oligopeptidase family protein	Metabolism of Cofactors and Vitamins pyrroloquinoline quinone biosynthesis	PP0375-0378
PP0376	pqqE	pyrroloquinoline quinone biosynthesis protein PqqE		
PP0377	pqqD	pyrroloquinoline quinone biosynthesis protein PqqD		
PP0378	pqqC	pyrroloquinoline quinone biosynthesis protein PqqC		
PP0379	pqqB	pyrroloquinoline quinone biosynthesis protein PqqB		
PP0380	pqqA	coenzyme PQQ synthesis protein PqqA		
PP0401	ksgA	dimethyladenosine transferase	???	PP0401-0403
PP0402	pdxA	4-hydroxythreonine-4-phosphate dehydrogenase		
PP0403	surA	survival protein SurA		
PP0404	imp			
PP0415	rpe	ribulose-phosphate 3-epimerase	Carbohydrate transport and metabolism	PP0411-0417
PP0416	gph	phosphoglycolate phosphatase		
PP0417	trpE	anthranilate synthase component I		
PP_t07		tRNA-Thr	Translation, ribosomal structure and biogenesis	PP_t08-PP0442
PP0440	tuf	elongation factor Tu		
PP_t08		tRNA-Trp		
PP0441	secE	preprotein translocase subunit SecE		
PP0442	nusG	transcription antitermination protein		
PP0443	rplK	50S ribosomal protein L11		PP0443-0444
PP0444	rplA	50S ribosomal protein L1		
PP0445	rplJ	50S ribosomal protein L10		
PP0446	rplL	50S ribosomal protein L7/L12		

Table 3.11 continued

verified operons	gene name	function	functional category	current annotation[#]
PP0449	rpsL	30S ribosomal protein S12		
PP0450	rpsG	30S ribosomal protein S7		
PP0451	fusA	elongation factor G		PP0451-0452
PP0452	tuf	elongation factor Tu		
PP0468	rpsH	30S ribosomal protein S8	Translation, ribosomal structure and biogenesis	PP0468-0475
PP0469	rplF	50S ribosomal protein L6		
PP0470	rplR	50S ribosomal protein L18		
PP0471	rpsE	30S ribosomal protein S5		
PP0472	rpmD	50S ribosomal protein L30		
PP0473	rplO	50S ribosomal protein L15		
PP0474	secY	preprotein translocase subunit SecY		
PP0475	rpmJ	50S ribosomal protein L36		
PP0476	rpsM	30S ribosomal protein S13		
PP0477	rpsK	30S ribosomal protein S11		
PP0478	rpsD	30S ribosomal protein S4		
PP0479	rpoA	DNA-directed RNA polymerase subunit alpha		
PP0480	rplQ	50S ribosomal protein L17		
PP0519	thiL	thiamine monophosphate kinase		PP0517-0521
PP0520	pgpA	phosphatidylglycerophosphatase A		
PP0521		hypothetical		
PP0719		GTP-dependent nucleic acid-binding protein EngD	Translation, ribosomal structure and biogenesis	PP0719-0725
PP0720	pth	peptidyl-tRNA hydrolase		
PP0768		response regulator/TPR domain-containing protein		PP0768-0769
PP0769		histidine kinase		
PP0770		PemI-like protein		PP0770-0772
PP0771		toxin ChpB		
PP0772		beta-lactamase domain protein		
PP0809	dsbB	disulfide bond formation protein B	Energy production and conversion	PP0809-0816
PP0810	cyoups1	cyoups1 protein		
PP0811	cyoups2	cyoups2 protein		
PP0812	cyoA	ubiquinol oxidase subunit 2		PP0809-0816
PP0813	cyoB	cytochrome o ubiquinol oxidase, subunit I		
PP0814	cyoC	cytochrome o ubiquinol oxidase, subunit III		
PP0815	cyoD	cytochrome o ubiquinol oxidase, protein CyoD		
PP0816	cyoE-2	protoheme IX farnesyltransferase		
[1]PP0833	tgt	queuine tRNA-ribosyltransferase	Intracellular trafficking and secretion	PP0832-0836
PP0834	yajC	preprotein translocase subunit YajC		
PP0835	secD	preprotein translocase subunit SecD		
PP0836	secF	preprotein translocase subunit SecF		
[2]PP0841		BadM/Rrf2 family transcriptional regulator		PP0841-0848
PP0842	iscS	cysteine desulfurase		
PP0843	iscU	scaffold protein		
PP0844	iscA	iron-sulfur cluster assembly protein IscA		
PP0845	hscB	co-chaperone HscB		
PP0846	hscA	chaperone protein HscA		
PP0847		ferredoxin, 2Fe-2S type, ISC system		
PP0848		hypothetical protein		
PP0849	ndk	nucleoside-diphosphate kinase		
PP0850		radical SAM enzyme, Cfr family		
PP0851	pilF	type IV pilus biogenesis/stability protein PilW		PP0849-0856
PP0852		Cro/CI family transcriptional regulator		
PP0853	ispG	4-hydroxy-3-methylbut-2-en-1-yl diphosphate synthase		
PP0854	hisS	histidyl-tRNA synthetase		
PP0855		hypothetical protein		
PP0856		hypothetical protein		

[1] The current operon prediction is incorrect. The four genes constitute either one operon or two operons, containing PP0833-0834 or PP0835-0836 respectively.

[2] The current operon prediction is incorrect. The four genes constitute either one operon or two operons, containing PP0841-0846 or PP0847-0850 respectively.

Table 3.11 continued

verified operons	gene name	function	functional category	current annotation[#]
PP0955		hypothetical protein		PP0953-0957
PP0956		YrbI family phosphatase		
PP0957		KpsF/GutQ family protein		
PP1068		amino acid ABC transporter ATP-binding protein	ABC transporter system	PP1065-1070
PP1069		polar amino acid ABC transporter inner membrane subunit		
PP1070		polar amino acid ABC transporter inner membrane subunit		
PP1352		putative nucleotide-binding protein		no prediction
PP1353		MscS mechanosensitive ion channel		
PP1431	lepA	GTP-binding protein LepA	Cell envelope biogenesis, outer membrane	PP1431-1432
PP1432	lepB	signal peptidase I		
PP1433	rnc	ribonuclease III		PP1433-1434
PP1434	era	GTP-binding protein Era		
PP1689		aromatic hydrocarbon degradation protein		no prediction
PP1690		hypothetical protein		
PP1691		hypothetical protein		
PP1797		HlyD family secretion protein		
PP1798		outer membrane efflux protein		
PP1910		hypothetical protein	Translation, ribosomal structure and biogenesis	PP1910-1912
PP1911	rpmF	50S ribosomal protein L32		
PP1912	plsX	putative glycerol-3-phosphate acyltransferase PlsX		
PP1913	fabD	malonyl CoA-acyl carrier protein transacylase	Fatty acid biosynthesis	PP1913-1914
PP1935.1		hypothetical protein		
PP1935.2		hypothetical protein		
PP1935.3		hypothetical protein		
PP1935.4		hypothetical protein		
PP1941		hypothetical protein		no prediction
PP1942		LysR family transcriptional regulator		
PP1960		hypothetical protein		no prediction
PP1961		hypothetical protein		
PP1962		phage integrase family site specific recombinase		PP1961-1964
PP1963		hypothetical protein		
PP3104		hypothetical protein		PP3105-3106
PP3105		hypothetical protein		
PP3106		hypothetical protein		
PP3896		Cro/CI family transcriptional regulator		PP3896-3898
PP3897		hypothetical protein		
PP3998		hypothetical protein		PP3896-3898
PP3898.1		hypothetical protein		
PP3898.2		hypothetical protein		
PP3987.1		hypothetical protein		
PP3988		hypothetical protein		
PP3989		DNA-cytosine methyltransferase		
PP4338	cheA	CheA signal transduction histidine kinase	Signal tranduction	PP4339-4340
PP4339	cheZ	chemotaxis phosphatase, CheZ		
PP4340	cheY	response regulator receiver protein		
PP4348		cystathionine beta-lyase, putative	Cysteine metabolism	PP4347-4348
PP4349		hypothetical protein		
PP4350		aminotransferase		
PP4374		hypothetical protein	Cell motility	PP4374-4375
PP4375	fliS	flagellar protein FliS		
PP4376	fliD	flagellar cap protein FliD		
PP4377		flagellin FlaG, putative		PP4377-4378
PP4378	fliC	flagellin FliC		
PP4388	flgE	flagellar hook protein FlgE	Cell motility	PP4387-4391
PP4389	flgD	flagellar basal body rod modification protein		
PP4390	flgC	flagellar basal body rod protein FlgC		
PP4391	flgB	flagellar basal body rod protein FlgB		

Table 3.11 continued

verified operons	gene name	function	functional category	current annotation[#]
PP4395	flgM	anti-sigma-28 factor, FlgM		PP4395-4396
PP4396		FlgN family protein		
PP4397		type IV pilus assembly PilZ		
PP4716	tpiA	triosephosphate isomerase		PP4715-4718
PP4717	glmM	phosphoglucosamine mutase		
PP4718	folP	dihydropteroate synthase		
PP4719	ftsH	ATP-dependent metalloprotease FtsH		
PP4740	hsdR	type I restriction-modification system, R subunit		PP4741-4742
PP4741	hsdM	type I restriction-modification system, M subunit		
PP4742	hsdS	type I restriction-modification system, S subunit		
PP4742.1		hypothetical		
PP4794	leuS	leucyl-tRNA synthetase	Translation, ribosomal structure and biogenesis	PP4794-4796
PP4795		rare lipoprotein B		
PP4800	lipA	lipoyl synthase	Coenzyme metabolism	PP4800-4802
PP4801	lipB	lipoyltransferase		
PP4802		hypothetical protein		
PP4803	dacA	Serine-type D-Ala-D-Ala carboxypeptidase		
PP4807	rodA	rod shape-determining protein RodA		PP4804-4811
PP4808		rRNA large subunit methyltransferase		
PP4809		iojap-like protein		
PP4810	nadD	nicotinic acid mononucleotide adenylyltransferase		
PP4811	proA	gamma-glutamyl phosphate reductase		
PP4891	hflC	HflC protein	Posttranslational modification, protein turnover, chaperones	PP4891-4899
PP4892	hflK	HflK protein		
PP4893	hflX	GTP-binding protein HflX		
PP4894	hfq	RNA-binding protein Hfq		
PP4941		LmbE family protein		PP4938-4944
PP4942		Mig-14 family protein		
PP4943		glycosyl transferase, putative		
PP4976	ahcY			no prediction
PP4977	metF	5,10-methylenetetrahydrofolate reductase		
PP4995	algH	hypothetical protein		PP4995-4999
PP4996		Holliday junction resolvase-like protein		
PP5044		GTP-binding protein TypA	Amino acid metabolism	no prediction
PP5045		thiamine biosynthesis protein ThiI	thiamine biosynthesis	
PP5075	gltD	glutamate synthase subunit beta	Amino acid metabolism	no prediction
PP5076	gltB	glutamate synthase subunit alpha	Alanine, aspartate and glutamate metabolism	
PP5281	rpmG	50S ribosomal protein L33	Protein synthesis	PP5280-5282
PP5282	rpmB	50S ribosomal protein L28		
PP5404		transposition protein, TnsC-related protein		PP5404-5405
PP5405		transposase, TnsB-related protein		
PP5406		transposase, TnsA-related protein		

Discussion

The main goal of the transcriptome analysis based on cDNA sequencing was the identification of expressed intergenic regions in *P. putida* KT2440, as these are generally not represented on commercially available microarrays but may contain so far unidentified genes, especially short open reading frames or sRNA genes. As such deep sequencing provides a hypothesis-free study of gene expression.

A complete reannotation of the *P. putida* KT2440 genome was not feasible, as cDNA sequencing data was available for two samples only. These samples represented the transcriptome status during growth under standard conditions and in response to cold shock and consequently did not contain transcripts from all gene loci. This, together with transcripts differing in length for some genes from the current annotation and identification of sRNA genes not described so far for *P. putida* demonstrated that in the initial annotation (Nelson *et al.*, 2002), many potential protein coding ORFs and regulatory sRNAs had been missed.

Since 2002, when the *P. putida* KT2440 genome was first sequenced, sequencing technology has developed dramatically. The first bacterial genomes were sequenced with the Sanger technology: *Haemophilus influenzae* (Fleischmann *et al.*, 1995), *Escherichia coli* (Blattner *et al.*, 1997) and *Mycobacterium tuberculosis* (Cole *et al.*, 1998) applying shotgun Sanger sequencing. The Roche/454 sequencing technology has enabled massively sequencing of whole genomes, saving time and costs (Margulies *et al.*, 2005).

However, the introduction of further next generation sequencing technologies such as Illumina and SOLiD opened a new era in genome research by further reducing costs and increasing throughput. Illumina and SOLiD sequencing can generate up to 17 Gb per week and 10-15 Gb per 3-7 days, respectively with costs of about $6 per Mb, in contrast to 454 sequencing generating about 500 Mb per 10 hours for more than $80 per Mb (Tucker *et al.*, 2009). Though the short read length (36 – 75 bp) provided by the new generation sequencing technologies make *de novo* assembly of novel genomes difficult, it is a useful approach for comparative genomics as indicated by a previous study by Tettelin *et al.* (2005), which was still performed with the whole genome shotgun Sanger sequencing approach. This study revealed by sequencing eight isolates of *Streptococcus agalactiae* strains representing the five major disease-causing serotypes that each sequenced genome contained many novel genes, demonstrating the genetic diversity even within one species or strain.

Recent studies have demonstrated that transcriptome analysis using cDNA sequencing (RNA-Seq) gives new insights into the genome-wide transcriptional organization and regulatory mechanisms, by revealing transcriptional start sites within operons, noncoding RNAs and antisense activity (Oliver et al., 2009; Filiatraut et al., 2010; Sharma et al., 2010). These features are all generally overlooked with microarray technologies.

In a comparative study of *Listeria monocytogenes* 10403S wild type strain and an isogenic mutant $\Delta sigB$ lacking the general stress response sigma factor σ^B Oliver et al., (2009) analyzed the transcriptome of stationary phase grown cells to reveal stress responsive genes that show σ^B-dependent transcription when entering stationary phase. In total they found 96 σ^B-dependent genes indicated by higher transcript levels in the wild type strain compared to the isogenic mutant. Furthermore, they could identify 65 putative σ^B promoters regions and by RNA-Seq analysis found 67 potential sRNAs transcribed in stationary phase, of which seven were previously unpredicted.

In a novel approach (dRNA-Seq) Sharma et al., 2010 selected for the 5' end of primary transcripts of *Helicobacter pylori* and by this identified 337 primary operons and 1907 transcriptional start sites (with some start sites even within operons) and antisense to annotated genes. The analysis of antisense transcription revealed that about 27% of the primary transcription start sites are also antisense, which indicates that antisense transcription occurs over the entire genome and additionally that the transcriptome is highly complex. They also discovered a high number of approximately 60 small RNAs that were all verified by Northern blot analysis.

Similarly, Filiatrault et al. (2010) used a strand-specific sequencing approach for enriched mRNA of *Pseudomonas syringae* DC3000, where they were able to identify 124 genes with sense and antisense expression at the same time. Furthermore, they detected 106 expressed regions inconsistent with the current annotation, of which 11 were incorporated into a new Genbank annotation record. Additionally, based on consistent expression they found 21 small RNAs.

For the transcriptome analysis of *P. putida* KT2440 in response to cold shock, non-enriched cDNA samples were sequenced using 36 bp reads from the Illumina Genome Analyzer. Beside the expression profile under standard growth condition and gene regulation in response to cold shock, the analysis of intergenic regions revealed 143 protein coding open reading frames located in 105 intergenic regions missed in the previous annotation and unclassified transcript that could not be assigned to protein or small RNA coding regions. Furthermore, 16 sRNA genes could be verified due to their consistent expression. These data correlate very well with the previously reported studies that revealed a similar number of newly identified ORFs, sRNAs and otherwise expressed intergenic regions. These studies demonstrate that many more protein coding ORFs and sRNAs are encoded on the genome than predicted in the initial annotation and might have been overlooked due to their small size or difficulties in predicting an open reading frame respectively. Sequencing of bacterial transcriptomes improves the current annotation and furthermore enables the genome-wide analysis of expression profiles, helping to better understand transcriptional regulatory networks.

Though transcriptome sequencing facilitates the identification of new transcripts, little is known about the cellular role of small proteins and sRNAs. However, those sRNAs with known function often play important regulatory roles in signal transduction pathways (Lapouge et al., 2007; Sonnleitner et al., 2009; Waters & Storz, 2009). Small proteins have been described to act as intracellular toxins (Fozo et al., 2008) and antibiotics (Kleerebezem, 2004).
Recently, two genome-based approaches for revealing the function of small proteins and sRNAs in E. coli have been published (Hobbs et al., 2010; Hemm et al., 2010). So far, the E. coli genome is known to encode more than 80 sRNAs and approximately 60 small proteins less than 50 amino acids long (Hemm et al., 2008; Waters & Storz, 2009).
With the generation and screening of 116 bar-coded deletion mutants affected in genes encoding small proteins and sRNAs for acid shock or cell envelope stress sensitive phenotypes, Hobbs et al. (2010) identified 14 sensitive mutants. Nine were sensitive to cell envelope stress and four to acid shock, while another two mutants were resistant to cell envelope stress. None of these had previously been associated with cell envelope stress.
Hemm et al. (2010) analysed the accumulation of SPA-tagged small proteins under different growth conditions and in response to stress and thus identified a number of small proteins that are synthesized only under specific conditions. One protein, YobF, was even demonstrated to exhibit a heat shock response at a posttranslational level.
The detection of many sRNAs during the last few years and in conjunction with the revelation of their regulatory roles in different cellular processes has put a major focus on RNA research. This is exemplified by Current Opinion in Microbiology addressing two issues to this topic (issues 2 and 3, 2007). Beside the known impact on mRNA stability, transcriptional and translational efficiency, recent studies showed the broad range of regulatory functions of small RNAs involved in processes such as quorum sensing (Bejerano-Sagie & Xavier, 2007), pathogenesis (Toledo-Arana et al., 2007), iron homeostasis (Massé et al., 2007), sugar metabolism (Vanderpool, 2007) and outer membrane protein synthesis (Valentin-Hansen et al., 2007) or very recently in carbon catabolite repression in P. aeruginosa (Sonnleitner et al., 2009).

Three sRNAs verified in this study were significantly differentially expressed in response to cold shock.

CrcZ is located downstream of the CbrAB two-component system, which is a conserved gene arrangement among fluorescent *Pseudomonads*. CrcZ was 7.8-fold down-regulated after cold shock, and its expression has previously been demonstrated to be CbrB-dependent (Sonnleitner *et al.*, 2009). Similarly to the Gac/Rsm signal transduction cascade (Lapouge *et al.*, 2007, 2008), the small RNA CrcZ sequesters the mRNA binding protein Crc. By this mechanism CrcZ regulates carbon catabolite repression, as low levels lead to repression and high levels to derepression (Sonnleitner *et al.*, 2009). The down-regulation of CrcZ in response to cold-shock in combination with the cold-sensitive phenotypes of the CbrAB two-component system and the PcnB poly(A) polymerase suggest that the expression level of CrcZ is also an important feature in adaptation processes to low temperatures.

The TPP (thiamin pyrophosphate) riboswitch is a highly conserved RNA structure among all three kingdoms of life. This riboswitch regulates gene expression by directly binding to TPP due to conformational changes, which leads either to transcription termination or inhibition of translation initiation (Miranda-Ríos, 2007). Thiamin pyrophosphate is the biologically active form of the vitamin B_1 and an essential cofactor for enzymes such as pyruvate dehydrogenase and 2-oxoglutarate dehydrogenase (Pohl *et al.*, 2004) which are involved in the tricarboxylic acid (TCA) cycle. The TCA cycle produces precursors for essential biosynthetic pathways such as purines and pyrimidines, amino acids and vitamins. Since TPP is essential for pyruvate dehydrogenase as it supplies acetyl-CoA to the TCA cycle and for 2-oxoglutarate dehydrogenase by the transformation of 2-oxoglutarate to succinyl-CoA, low levels of TPP impair the TCA cycle (Dunn, 1998). The TPP riboswitch was highly down-regulated (9.7-fold) reflecting a general reduced energy demand at low temperatures and in this context reduction of oxidative phosphorylation and ATP synthesis.

The sRNA P1 was predicted and verified in *P. aeruginosa* and is at least conserved among *Pseudomonas* species where it is located upstream of a glutamine synthetase gene (Livny *et al.*, 2006; Filiatraut *et al.*, 2010). Though no regulatory function has been predicted so far, the significant up-regulation (4.7-fold) indicates a direct response of this sRNA to cold shock.

Though experimental validation of identified transcripts is still pending to determine if transcripts code for proteins or sRNAs, the sequencing of the *P. putida* KT2440 transcriptome demonstrates that many more genes and putative regulatory sRNAs are encoded in the genome than previously thought and that the majority of these might play a direct role in the cold shock response. This is indicated by the over-represented number of up-regulated genes and high expression in intergenic regions. Astonishing, 87% of the newly predicted protein coding ORFs and 75% of the unclassified transcripts were up-regulated after cold shock.

This study, together with the previously described studies (Oliver *et al.*, 2009; Filiatraut *et al.*, 2010; Sharma *et al.*, 2010), highlights the potential and benefits of deep RNA sequencing, provides new strategies for genome-wide transcriptome analysis up to single basepair resolution and hence enables the exploration of transcriptional regulation in more detail and may even reveal previously unrecognized regulatory networks. Deep RNA sequencing does suffer from technical biases as well, e.g. in the PCR based amplification steps of cDNA fragments and when sequencing AT-rich genomes (Kozarewa *et al.*, 2009). The advantages over gene expression microarrays, however, are significant as already indicated in this analysis. The advantages and disadvantages of deep RNA sequencing in comparison to microarrays are presented and discussed in more detail in the next chapter (3.5) where two different microarrays and the deep RNA sequencing technology are compared.

3.6 Comparison of Three Transcriptome Platforms

The objective of this study was the detailed analysis of stress response in *P. putida* KT2440, mainly based on transcriptome analysis. For this, three different transcriptome platforms were included in the study, two different gene expression microarrays (Progenika and Affymetrix, see 2.1.2.1 and 2.1.2.2) and the Illumina cDNA sequencing technology. The same RNA samples derived either from 30°C or 10°C cultures were transcribed into cDNA and analyzed by the three methods in order to compare the three transcriptome platforms.

One major limitation of microarrays in general is that only predefined sequences can be detected. In this case, the analysis of expression in intergenic regions is limited as sequences representative for intergenic regions are only weakly covered on microarrays that are available for the *P. putida* genome. Affymetrix microarrays do indeed include sequences representing 2443 intergenic regions, but still this is only about half of all intergenic regions of the *P. putida* KT2440 genome. Progenika microarrays cannot detect intergenic regions at all, as these are not spotted on the array due to the selection of oligonucleotide. The focus on analyzing the expression profile of predefined sequences rules out the identification of novel undescribed transcripts, such as ORFs and sRNAs. This capability of sequenced based analysis was demonstrated in the previous chapter by examining the expression of intergenic regions.

For this reason, the three transcriptome platforms were solely compared based on the detected signal intensity of annotated genes, as published by Nelson *et al.* (2002). Mean and median values of signal intensities were calculated for the microarrays and the normalized read numbers (RPKM, reads per kilobase and million) for the sequencing output. Subsequently, a rank number evaluation was performed to facilitate platform. Signal intensities and RPKM values in case of the sequencing output corresponding to the 5420 predicted ORFs from the initial annotation were attributed with rank numbers. The highest signal intensity (or RPKM) was assigned the rank number 1 and the lowest 5420. In case of equal values of signal intensity (or RPKM) rank numbers were allocated in ascending order to the ORF numbers, thus no rank number was allocated twice.

For the transcriptome analysis of cold shock response with the Progenika microarrays, two biological samples each for wild type and respective mutants were examined. However, initial analysis of the two biological wild type replicates implicated failures occurred during sample preparation of the second biological sample since the transcriptome raw data were not evaluable. As statistical analysis indicating no differential expression with statistical reliability and the overall poor signal intensities, the second sample was excluded from further analysis. Retrospectively, it is difficult to determine the error that caused erroneous values for this sample. However, as cDNA generation and labelling were performed simultaneously, it might be that the hybridization step was insufficient, leading to the weak and inconsistent signal intensities. However, for the transcriptome analysis with Affymetrix and Illumina, cDNA was only prepared from the first biological sample due to the limitations of available Affymetrix microarrays and lanes for Illumina cDNA sequencing.

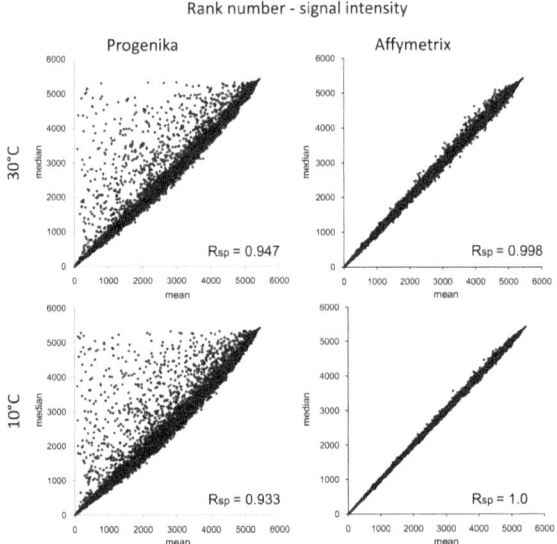

Fig. 3.19 Rank number evaluation of signal intensities for Progenika and Affymetrix microarrays. Rank numbers were assigned to mean and median values obtained from either Progenika or Affymetrix microarrays for the two tested conditions. Spearman correlation factor was calculated demonstrating a higher scatter of signal intensities detected with the Progenika microarrays.

RESULTS AND DISCUSSION

3.6.1 Correlation between the results from three transcriptome platforms

First, the two gene expression microarrays were compared by mean and median signal intensity. Rank numbers for mean and median signal intensity were assigned as described above and plotted against each other. Furthermore, the Spearman correlation factor was calculated.

As shown in Figure 3.19 the correlation of mean and median value of Affymetrix signal intensities was nearly 1 demonstrating a consistent distribution of signal intensities among the different Affymetrix microarrays. In contrast, the comparison of mean and median signal intensities derived from the Progenika microarrays showed strong scattering towards the median values. Though, the Spearman correlation was still good (0.947 for 30°C and 0.933 for 10°C), the scatter indicated that in a few cases the signal intensities varied strongly among the four different microarrays.

The design of the two microarrays is very different. The Progenika microarray is also a two-dye chip, and a major difference is caused by the distribution and abundance of oligonucleotide spots representing the different ORFs. On the Progenika microarray only one oligo per ORF is spotted twice, whereas in total 26 oligos (13 perfect match and 13 mismatch oligos) represent one ORF on the Affymetrix microarray. Furthermore, the two spots per ORFs on the Progenika microarray are spotted next to each other, in contrast to the Affymetrix microarrays, where they are equally distributed over the entire array. This provides a statistical more reliable allocation of signal intensities and can compensate for outliers, in case of locally poor or inconsistent hybridization results.

Based on this result, the mean signal intensity was chosen for further rank number evaluations.

The next step was to compare the different platforms. First, rank numbers of mean signal intensities and RPKM for 30°C and 10°C respectively were plotted against each other. All platforms were compared against each another and Spearman correlation factor was again calculated (see Figure 3.20, A). Secondly, the platforms were compared by plotting the rank number assigned to the fold-change against each other (see Figure 3.20, B).

Generally, the correlation between the platforms was rather poor independent of whether rank numbers of signal intensities or fold-changes were compared. Interestingly, the correlation was best between the Affymetrix and Illumina platforms with a Spearman correlation factor of about 0.74 when comparing rank numbers of signal intensities. The correlation between the Progenika microarray and either the Affymetrix microarray or the Illumina sequencing, however was poor with a correlation factor of about 0.53 for the two microarrays and 0.43 (30°C) or 0.45 (10°C) for the comparison of the Progenika and the Illumina platform.

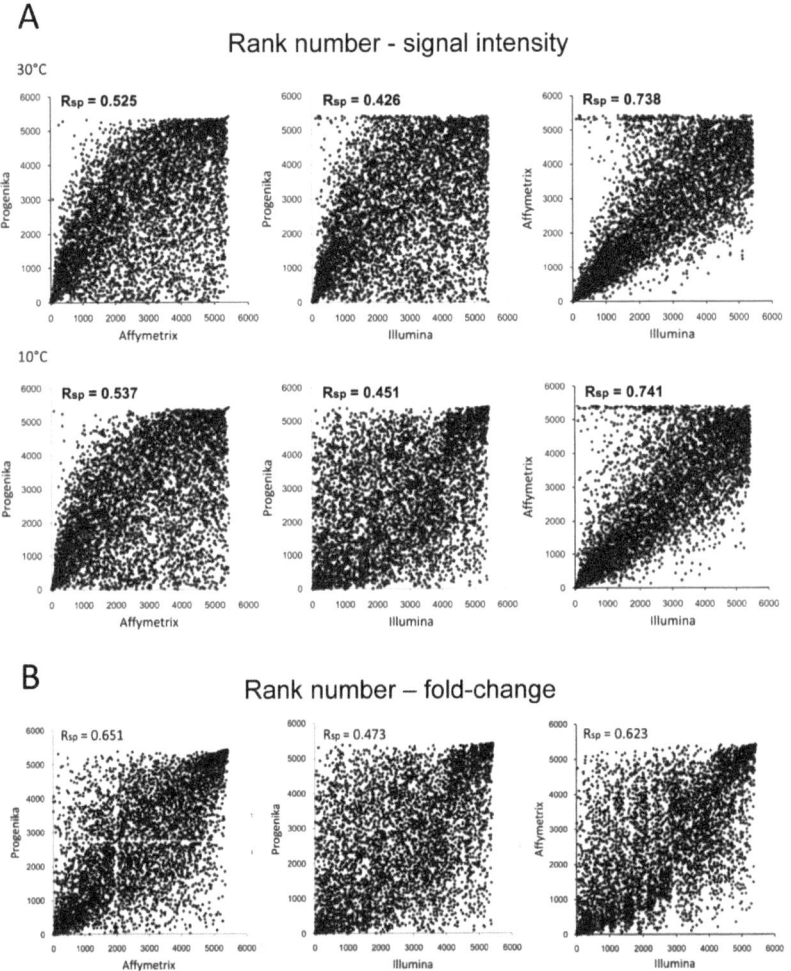

Fig. 3.20 Comparison of the three transcriptome platforms due to rank number assignment. **A)** Rank numbers were assigned either to signal intensities or RPKM (reads per kilobase and million) values obtained from the two microarray platforms or the Illumina cDNA sequencing technology respectively at 30°C and 10°C. **B)** Rank numbers were assigned to fold-changes ascending from highest up-regulated to highest down-regulated gene. Spearman correlation factors show that the correlation among the three platforms is best for the comparison of Affymetrix and Illumina.

RESULTS AND DISCUSSION

Two features were apparent when examining the graphs.

ORFs with considerable coverage according to Illumina sequencing results were not detected with Affymetrix microarrays.
In Figure 3.20 (A) illustrating the correlation between Affymetrix and Illumina a line of dots was visible at the top of the scatter-plots, corresponding to genes attributed high rank numbers for Affymetrix and low for Illumina. Though the Spearman correlation between the two platforms was satisfactory (0.74), these spots indicated that ORFs with a medium expression level detected by cDNA sequencing were not detected with the Affymetrix microarrays. ORFs with an assigned rank number lower than 3000 and higher than 5200, according to Illumina and Affymetrix results respectively, were considered to be reliably expressed but not detected with Affymetrix microarrays. This group of ORFs was examined in more detail and is listed in Table 3.12.

In total 126 genes were identified to be sufficiently expressed, 60 of them under both tested conditions, according to the cDNA sequencing data, but were not detected with the Affymetrix microarrays. Looking in more detail into the Affymetrix design, 90 of the 5420 initially annotated ORFs (Nelson *et al.*, 2002) were not represented on the Affymetrix microarrays at all. Seventy-four of them were identified to be sufficiently expressed in the cDNA sequencing. Based on identified transcripts with sufficient and continuous coverage, for eight out of the 74 loci new protein coding open reading frames could be predicted, three of which encoded two proteins. All ten ORFs showed high homology to genes present in genomes of other *Pseudomonas* species.

In the most up to date annotation as provided by the *Pseudomonas* database (www.pseudomonas.com) 5350 protein coding ORFs are predicted, another 58 genes that were initially predicted to be protein coding were classified as genes, which do not apparently code for proteins due to frame-shift mutations (e.g. PP0253, PP4496) and another 12 genes were completely removed from the original annotation list. Out of these 70 genes, the majority were found to be sufficiently expressed under at least one condition. Eleven of the 12 removed genes and 41 of the 58 genes predicted to be non protein coding were transcribed and can be found in Table 3.12. For eight genes that were in the group of genes but non protein coding, a new open reading frame was predicted due to the identified transcript (see Table 3.13).

This result clearly demonstrates that the current *P. putida* KT2440 annotation needs to be revised as the majority of the 90 genes (82%) that were not spotted due to incorrect predictions were indeed sufficiently expressed when analyzing the cDNA sequencing data.

RESULTS AND DISCUSSION

The remaining 52 ORFs, which were spotted on the Affymetrix microarrays but not detected, were examined for conspicuous physicochemical features that might explain the weak hybridization. Though no secondary structures such as palindromes were detected, the analysis revealed that the genes were on average rather small with a gene length of 93-2364 bp, 153-840 bp, 273 bp (smallest to largest gene, inner percentile, median). The majority of genes (37) that were not detected with Affymetrix microarrays but were sufficiently expressed according to cDNA sequencing, were small genes coding for hypothetical proteins with a median gene length of 171 bp (see Table 3.15). Shorter ORFs make a statistical distribution of representative perfect and mismatch oligomers more difficult, and thus might cause a bias in the calculation of signal intensities. Shorter ORFs make a statistical distribution of the 26 representative perfect and mismatch oligonucleotides more difficult due to limited sequence space. Another striking feature was that the median GC-content of 54.3% of these ORFs was significantly below the chromosomal average (61.5%). Relatively AT-rich fragments might show a weaker hybridization affinity due to a non optimal hybridization temperature applied in the respective protocol. This temperature is calculated for the average GC-content of the genome and is therefore higher than the optimal hybridization temperature of these relatively AT-rich fragments. No significant difference in oligonucleotide parameters of the gene sequence could be detected using the OligoCounter (Ganesan et al., 2008; http://www.bi.up.ac.za/SeqWord/mhhapplet.php) and Oligowords (Davenport et al., 2009; http://webhost1.mh-hannover.de/davenport/oligocounter/index.html) programs.

RESULTS AND DISCUSSION

Table 3.12 Summary of genes that were not detected with Affymetrix array but showed sufficient expression with Illumina cDNA sequencing. All these genes have been assigned rank numbers ≥5200 for Affymetrix signal intensities and rank numbers ≤ 3000 for Illumina RPKM values.

#	locus id	Affymetrix - signal intensity				Illumina - RPKM				fold-change
		mean 30°C	RN 30°C	mean 10°C	RN 10°C	30°C	RN 30°C	10°C	RN 10°C	
1	PP0025	9.5292	5314	7.8415	5322	23.57	2131	19.36	2784	-1.22
1	PP0028	9.4031	5315	8.2359	5318	43.97	1452	18.06	2883	-2.44
1	PP0042	9.8337	5312	10.5809	5248	16.90	2549	13.88	3232	-1.22
1	PP0048	9.1995	5319	12.0776	5129	33.28	1743	63.78	1348	1.92
0	PP0156	0.0000	5324	0.0000	5324	121.34	721	48.81	1651	-2.49
0	PP0212	0.0000	5325	0.0000	5325	26.35	2001	21.64	2641	-1.22
0	PP0253	0.0000	5326	0.0000	5326	1334.10	112	133.86	703	-9.97
1	PP0374	12.6285	5232	138.0305	1772	181.43	544	298.04	326	1.64
0	PP0404	0.0000	5327	0.0000	5327	429.32	262	299.42	325	-1.43
0	PP0499	0.0000	5329	0.0000	5329	18.14	2466	18.63	2838	1.03
0	PP0511	0.0000	5330	0.0000	5330	10.87	3156	26.80	2360	2.47
0	PP0598	0.0000	5331	0.0000	5331	25.41	2044	310.78	312	12.23
0	PP0623	0.0000	5332	0.0000	5332	212.31	477	361.00	268	1.70
1	PP0731	11.6524	5279	10.7510	5237	5.81	4066	28.62	2283	4.93
1	PP0783	12.2511	5246	12.7639	5066	19.69	2361	113.20	836	5.75
0	PP1158	0.0000	5333	0.0000	5333	4.00	4617	144.69	651	36.17
1	PP1241	10.9967	5296	10.8729	5229	20.48	2320	8.41	3892	-2.44
0	PP1327	0.0000	5334	0.0000	5334	580.14	201	946.00	126	1.63
0	PP1409	0.0000	5335	0.0000	5335	5.18	4226	72.30	1231	13.96
1	PP1473	12.8351	5220	16.5799	4683	37.07	1621	10.15	3633	-3.65
1	PP1485	11.6774	5278	9.7098	5287	26.74	1984	21.96	2623	-1.22
0	PP1495	0.0000	5336	0.0000	5336	513.76	224	206.96	455	-2.48
1	PP1687	8.9890	5320	9.6491	5292	11.98	2977	39.37	1891	3.29
1	PP1724	16.2770	5027	10.9316	5223	4.27	4506	17.54	2922	4.11
0	PP1731	0.0000	5337	0.0000	5337	22.57	2195	41.71	1817	1.85
0	PP1796	0.0000	5338	0.0000	5338	47.78	1374	647.48	159	13.55
1	PP1848	10.6684	5302	7.9617	5320	31.71	1797	26.05	2390	-1.22
1	PP1856	11.3212	5285	8.4013	5317	34.45	1698	28.30	2293	-1.22
0	PP1906	0.0000	5339	0.0000	5339	41.23	1505	83.12	1095	2.02
0	PP1919	0.0000	5340	0.0000	5340	77.90	1003	79.98	1136	1.03
0	PP1939	0.0000	5341	0.0000	5341	25.06	2070	20.58	2714	-1.22
1	PP1943	17.1856	4953	10.5733	5250	8.33	3557	27.38	2336	3.29
0	PP1965	0.0000	5342	0.0000	5342	29.21	1888	95.97	967	3.29
1	PP1973	9.7774	5313	8.9658	5311	36.56	1629	30.03	2216	-1.22
0	PP1994	0.0000	5343	0.0000	5343	33.33	1739	30.80	2180	-1.08
0	PP2110	0.0000	5344	0.0000	5344	11.88	2992	29.27	2249	2.46
1	PP2221	11.6276	5280	14.9583	4839	21.08	2286	17.31	2936	-1.22
1	PP2321	15.2756	5085	11.0082	5216	37.72	1608	30.98	2170	-1.22
0	PP2391	0.0000	5345	0.0000	5345	14.63	2741	12.01	3445	-1.22
1	PP2520	11.9649	5263	11.8117	5154	30.89	1829	25.37	2425	-1.22
0	PP2521	0.0000	5347	0.0000	5347	23.09	2162	11.38	3507	-2.03
0	PP2523	0.0000	5348	0.0000	5348	20.02	2342	16.44	2999	-1.22
1	PP2571	9.3641	5317	8.4753	5316	33.49	1731	27.51	2322	-1.22
1	PP2670	12.9083	5215	10.4032	5257	75.21	1024	30.89	2173	-2.43
1	PP2763	10.5681	5305	9.3185	5304	23.12	2159	18.99	2815	-1.22
1	PP2810	13.8904	5169	9.7578	5282	2.52	5204	16.57	2988	6.58
0	PP2966	0.0000	5352	0.0000	5352	21.08	2287	69.25	1271	3.29
0	PP2978	0.0000	5354	0.0000	5354	8.94	3457	29.36	2245	3.28
0	PP2981	0.0000	5355	0.0000	5355	14.76	2732	36.37	1999	2.46
0	PP2982	0.0000	5356	0.0000	5356	3.59	4780	26.54	2372	7.39
0	PP2983	0.0000	5357	0.0000	5357	10.92	3150	40.38	1856	3.70
0	PP2984	0.0000	5358	0.0000	5358	29.92	1860	92.16	1003	3.08
1	PP2992	12.0571	5258	10.9808	5220	13.09	2863	21.51	2650	1.64
0	PP3113	0.0000	5359	0.0000	5359	9.98	3287	32.79	2106	3.29
0	PP3114	0.0000	5360	0.0000	5360	39.76	1555	24.97	2447	-1.59
0	PP3172	0.0000	5361	0.0000	5361	22.69	2183	66.28	1310	2.92
0	PP3334	0.0000	5365	0.0000	5365	12.62	2916	10.36	3616	-1.22
1	PP3351	9.8655	5310	8.6117	5315	33.49	1732	27.51	2323	-1.22
0	PP3365	0.0000	5366	0.0000	5366	205.00	495	164.80	557	-1.24
1	PP3373	9.9914	5309	9.0352	5310	137.99	657	72.24	1232	-1.91
0	PP3381	0.0000	5367	0.0000	5367	13.49	2829	59.84	1430	4.44
0	PP3399	0.0000	5368	0.0000	5368	9.92	3294	24.44	2476	2.46
0	PP3408	0.0000	5369	0.0000	5369	23.05	2165	32.46	2125	1.41
1	PP3442	10.9413	5298	8.7224	5314	21.46	2266	17.62	2914	-1.22

RESULTS AND DISCUSSION

Table 3.12 continued

#	locus id	Affymetrix - signal intensity				Illumina - RPKM				fold-change
		mean 30°C	RN 30°C	mean 10°C	RN 10°C	30°C	RN 30°C	10°C	RN 10°C	
0	PP3470	0.0000	5370	0.0000	5370	18.86	2411	46.47	1708	2.46
0	PP3586	0.0000	5372	0.0000	5372	2.70	5129	19.95	2752	7.39
1	PP3617	11.9385	5265	29.3731	3733	20.00	2345	65.69	1320	3.28
1	PP3619	10.7141	5300	15.4922	4777	12.00	2974	39.42	1889	3.28
0	PP3673	0.0000	5373	0.0000	5373	7.18	3805	17.69	2910	2.46
0	PP3674	0.0000	5374	0.0000	5374	10.07	3273	16.53	2991	1.64
1	PP3675	9.3393	5318	11.1309	5203	23.57	2134	19.36	2787	-1.22
1	PP3676	11.3600	5284	12.3637	5104	15.28	2678	46.01	1723	3.01
1	PP3679	11.7775	5273	9.3430	5303	11.09	3120	36.45	1992	3.29
0	PP3687	0.0000	5375	0.0000	5375	14.63	2743	12.01	3447	-1.22
1	PP3809	8.6273	5322	9.2608	5305	38.95	1571	31.99	2138	-1.22
0	PP3820	0.0000	5379	0.0000	5379	25.22	2057	91.13	1016	3.61
1	PP3846	14.1246	5155	10.7266	5238	8.79	3482	28.89	2266	3.29
0	PP3868	0.0000	5380	0.0000	5380	25.22	2058	51.78	1589	2.05
1	PP3872	11.9623	5264	9.4742	5299	30.11	1855	24.73	2463	-1.22
1	PP3885	10.6043	5304	12.5797	5088	301.39	360	357.57	271	1.19
1	PP3937	11.7613	5275	10.4985	5254	29.37	1882	24.12	2492	-1.22
1	PP3969	11.0708	5294	11.0608	5210	19.80	2353	32.52	2120	1.64
0	PP3984	0.0000	5382	0.0000	5382	27.56	1949	124.52	751	4.52
1	PP3987	11.1494	5291	11.1023	5206	7.47	3742	116.62	809	15.61
0	PP4023	0.0000	5383	0.0000	5383	33.38	1737	36.56	1986	1.10
0	PP4024	0.0000	5384	0.0000	5384	19.96	2347	49.19	1642	2.46
0	PP4025	0.0000	5385	0.0000	5385	140.33	649	165.21	555	1.18
0	PP4026	0.0000	5386	0.0000	5386	19.37	2389	143.18	659	7.39
0	PP4039	0.0000	5388	0.0000	5388	11.03	3132	27.17	2347	2.46
1	PP4062	13.8065	5174	9.8419	5278	36.56	1633	30.03	2218	-1.22
0	PP4082	0.0000	5390	0.0000	5390	20.87	2300	11.43	3501	-1.83
1	PP4087	9.8359	5311	71.2435	2582	23.57	2137	232.35	403	9.86
0	PP4088	0.0000	5392	0.0000	5392	2.43	5230	21.95	2627	9.03
0	PP4089	0.0000	5393	0.0000	5393	19.78	2354	64.99	1331	3.29
0	PP4090	0.0000	5394	0.0000	5394	15.79	2642	58.34	1451	3.69
0	PP4093	0.0000	5395	0.0000	5395	36.07	1644	44.44	1758	1.23
0	PP4140	0.0000	5397	0.0000	5397	19.11	2402	69.33	1268	3.63
1	PP4240	8.7864	5321	14.5932	4877	20.87	2301	48.97	1646	2.35
1	PP4241	7.4187	5323	7.5750	5323	15.79	2643	25.93	2398	1.64
1	PP4317	9.4026	5316	7.9868	5319	18.19	2464	14.94	3140	-1.22
0	PP4320	0.0000	5400	0.0000	5400	21.52	2259	28.28	2297	1.31
1	PP4348	11.0948	5293	81.1400	2432	13.10	2862	131.82	712	10.06
0	PP4351	0.0000	5401	0.0000	5401	13.62	2821	44.76	1749	3.29
1	PP4360	15.8032	5058	10.6162	5245	29.37	1883	24.12	2493	-1.22
1	PP4409	11.7573	5276	12.0755	5131	15.03	2698	14.41	3194	-1.04
1	PP4414	10.9776	5297	13.8177	4972	22.80	2177	40.13	1865	1.76
1	PP4428	12.3136	5240	13.3262	5020	25.17	2061	17.23	2943	-1.46
0	PP4436	0.0000	5403	0.0000	5403	16.44	2589	13.50	3288	-1.22
0	PP4440	0.0000	5404	0.0000	5404	7.43	3753	18.32	2862	2.47
0	PP4441	0.0000	5405	0.0000	5405	11.20	3105	18.39	2852	1.64
0	PP4443	0.0000	5407	0.0000	5407	14.01	2791	9.59	3707	-1.46
1	PP4453	13.6268	5182	9.7108	5286	39.12	1568	19.28	2793	-2.03
1	PP4488	13.7347	5178	10.3994	5258	14.11	2781	23.17	2542	1.64
0	PP4496	0.0000	5408	0.0000	5408	87.31	916	135.16	696	1.55
0	PP4536	0.0000	5409	0.0000	5409	11.79	3013	19.36	2789	1.64
0	PP4710	0.0000	5411	0.0000	5411	480.63	232	1839.04	83	3.83
0	PP4744	0.0000	5412	0.0000	5412	37.17	1620	146.54	645	3.94
0	PP4778	0.0000	5414	0.0000	5414	28.35	1921	32.60	2113	1.15
0	PP4878	0.0000	5415	0.0000	5415	16.00	2626	144.53	654	9.03
0	PP4884	0.0000	5416	0.0000	5416	24.05	2112	98.76	943	4.11
0	PP4976	0.0000	5417	0.0000	5417	596.77	195	177.47	517	-3.36
0	PP5013	0.0000	5418	0.0000	5418	39.79	1554	52.65	1576	1.32
1	PP5062	13.2553	5202	10.5749	5249	65.75	1122	102.60	913	1.56
1	PP5176	12.4895	5235	10.9927	5219	12.94	2888	15.94	3046	1.23
0	PP5290	0.0000	5419	0.0000	5419	48.27	1369	155.54	599	3.22
0	PP5394	0.0000	5420	0.0000	5420	14.13	2776	32.24	2132	2.28

\# 0: ORF not spotted on Affymetrix microarray, 1: ORF spotted on Affymetrix microarray.
black: Expression detected with Illumina cDNA sequencing under both temperature samples.
red: ORF was expressed at 30°C when analyzed with Illumina.
blue: ORF was expressed at 10°C when analyzed with Illumina.
bold: Genes show significant differential expression when analyzed with Illumina sequencing.

Table 3.13 Summary of newly annotated ORFs. The loci had been predicted to be non protein coding before, but showed sufficient expression in Illumina cDNA sequencing.

novel id	start	stop	length (bp)	Blast e-value	ortholoug	functional gene annotation
PP0253.1	307775	307248	528	3.00E-114	Pput_0268	Phosphoenolpyruvate carboxykinase (ATP)
PP0253.2	308511	307843	669	1.00E-38	Pput_0268	Phosphoenolpyruvate carboxykinase (ATP)
PP0404.1	489893	489597	297	0	Pput_0438	Organic solvent tolerance protein
PP0404.2	492094	489875	2220	3.00E-46	Pput_0438	Organic solvent tolerance protein
PP0598.1	703422	703919	498	1.00E-74	Pput_0638	Paraquat-inducible protein A
PP0598.2	703882	704529	648	7.00E-68	Pput_0639	Paraquat-inducible protein A
PP0623.1	728663	729472	810	2.00E-145	Pput_0663	RluA family pseudouridine synthase
*PP1495.1	1700805	1701551	747	7.00E-22	Pmen_3317	Peptide chain release factor 2
*PP1796.1	2017205	2018710	1506	0	PputW916_1405	Type I secretion system ATPase
PP4089.1	4623732	4624400	669	3.00E-30	PFLU6030	Hypothetical protein
PP4496.1	5111610	5111975	366	4.00E-08	PputGB1_4002	RNA-binding S4 domain protein

* ORF was already found to be misannotated previously (Table 3.8).

Table 3.14 Genes not detected with Affymetrix microarrays examined for gene length, GC-content and functional annotation.

locus id	gene length (bp)	GC-content (%)	functional annotation
PP0025	153	53.595	hypothetical protein
PP0028	327	47.706	hypothetical protein
PP0042	213	50.235	hypothetical protein
PP0048	324	48.457	hypothetical protein
PP0374	159	55.975	hypothetical protein
PP0731	618	55.663	phosphatidylserine synthase, putative
PP0783	183	48.634	hypothetical protein
PP1241	351	50.142	hypothetical protein
PP1473	291	60.825	hypothetical protein
PP1485	135	67.407	hypothetical protein
PP1687	300	57.333	hypothetical protein
PP1724	840	63.095	ABC transporter, permease protein
PP1848	114	59.649	hypothetical protein
PP1856	105	60.952	hypothetical protein
PP1943	861	49.593	formyltetrahydrofolate
PP1973	99	55.556	hypothetical protein
PP2221	171	53.216	hypothetical protein
PP2520	117	53.846	hypothetical protein
PP2571	108	58.333	hypothetical protein
PP2670	1335	66.142	hypothetical protein
PP2763	156	55.128	hypothetical protein
PP2810	1422	61.463	hypothetical protein
PP2992	822	56.569	hypothetical protein
PP3351	108	50.000	hypothetical protein
PP3373	2364	61.168	bacterial surface antigen family protein
PP3442	168	50.595	hypothetical protein
PP3617	897	53.066	hypothetical protein
PP3619	897	51.282	hypothetical protein
PP3675	153	43.137	hypothetical protein
PP3676	2112	42.756	hypothetical protein
PP3679	324	51.852	hypothetical protein
PP3809	93	52.688	hypothetical protein
PP3846	816	59.314	carbon-nitrogen hydrolase family protein
PP3872	120	65.000	hypothetical protein
PP3885	108	46.296	hypothetical protein
PP3937	123	54.472	hypothetical protein
PP3969	363	55.923	response regulator
PP3987	960	54.271	site-specific recombinase, phage integrase family
PP4062	99	54.545	hypothetical protein
PP4087	153	35.948	hypothetical protein
PP4240	1203	43.558	microcin b17 processing protein mcbd, putative
PP4241	228	47.807	hypothetical protein
PP4317	198	51.515	hypothetical protein
PP4348	1095	46.941	cystathionine beta-lyase, putative
PP4360	123	56.911	hypothetical protein
PP4409	1431	43.816	site-specific recombinase, phage integrase family
PP4414	93	54.839	hypothetical protein
PP4428	855	56.959	amino acid ABC transporter, periplasmic amino acid-binding protein
PP4452	1080	50.833	NAD/NADP octopine/nopaline dehydrogenase family protein
PP4488	255	56.863	hypothetical protein
PP5062	546	57.509	hypothetical protein
PP5176	555	55.135	hypothetical protein
Mean	513.92	53.741	
Median	273	54.372	

RESULTS AND DISCUSSION

Only few genes show a good correlation between Progenika and Affymetrix microarrays

The correlation of fold-changes between Progenika and Affymetrix exhibited a weak correlation with a Spearman factor of 0.65. Within the resulting cloud-like distribution, a cross-like pattern was visible caused by spots that were concentrated in the centre. The respective dots correlated to rank numbers of approximately 2700 for Progenika and 1950 for Affymetrix. These spots belonged to the 70 ORFs described above to be either non protein coding or were removed from the initial annotation and for this reason were not spotted on either of the two microarrays. As the rank numbers for the fold-change were assigned ascending from the most up-regulated gene (rank number 1) to the most down-regulated gene (rank number 5420), these 70 genes were all allocated with a fold-change of 1 simulating no change of expression, and thus resulting in a good correlation because they were consecutively ordered in the rank number list. This cross-like pattern appeared to be an artificial effect resulting from the assignment of rank numbers to all ORFs from the initial annotation. Since the respective ORFs were not spotted on the microarrays, they could not be detected and hence showed no expression. However, that only these 70 genes showed an artificially good correlation, demonstrates more over that even the correlation between the two microarray platforms is rather poor and the outcome is hardly comparable.

Other transcriptome studies have already mentioned the problem of the low comparability of different microarrays. For example, Pedotti *et al.* (2008) compared a total of five different microarray platforms that were later on again compared to deep RNA sequencing data that was derived from the same samples ('t Hoen *et al.*, 2008). The authors concluded that at least for their experimental setup where gene expression differences were expected to be marginal, different microarray platforms might give a complementary view on the biological processes. Furthermore, with the comparison of the microarray platforms and the deep RNA sequencing technology, they showed that the sequencing technology was more robust and comparable and had a more dynamic expression profile than the microarrays. This point will be discussed more extensively at the end of this chapter.

In the following sections the analysis and comparison of the three gene expression profiles is presented.

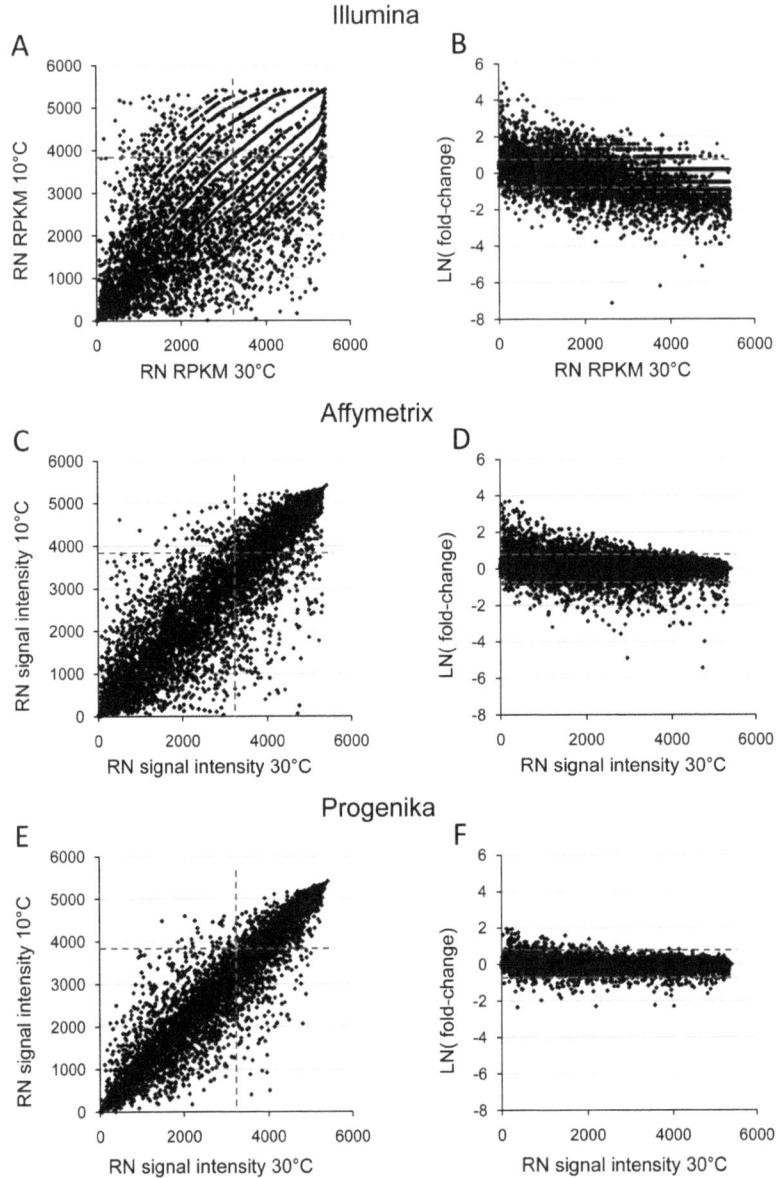

Fig. 3.21 Rank number evaluation of gene expression. Rank numbers were assigned ascending from highest to lowest expression level. **A, C, E)** Rank numbers of gene expression for 30°C and 10°C were spotted against one another. Dashed lines indicate rank numbers from which expression level is lower than 10 RPKM, always correlated to Illumina sequencing results. **B, D, F)** The natural logarithm of fold-change is spotted against rank numbers of gene expression at 30°C. Dashed lines indicate genes with differential expression higher than two-fold.

3.6.2 Gene expression profiles derived from the three transcriptome platforms

Subsequent to the rank number analysis, the three transcriptome platforms were compared based on the determined expression profiles. For this, firstly rank numbers according to the respective signal intensities for the 30°C and 10°C samples were plotted against each other separately for each platform (see Figure 3.21 A, C, E), and secondly the fold-changes on a logarithmic scale against rank numbers of signal intensities from the 30°C samples (see Figure 3.21 B, D, F).

The dotted lines in graphs A, C and E indicate the threshold for rank numbers representing low abundant genes (RPKM ≤10) detected with cDNA sequencing. The dotted line is in all three graphs representative for rank numbers assigned to Illumina RPKM values. In graphs B, D and E the dotted lines indicate the threshold for genes that are more than two-fold up or down-regulated.

The graphs demonstrated that the expression profile detected with cDNA sequencing was more dynamic than the two microarrays as the overall scatter was much stronger (A) and the absolute fold-change was much higher (B).

Regarding the absolute fold-changes, hardly any genes showed a differential expression in excess of 10-fold (ln(10) = 2.3) with the Progenika microarrays. This result was in direct contrast to the Affymetrix microarrays where many genes were detected to be differentially expressed in the range of 10 to 50-fold (ln(50) = 3.9). With the cDNA sequencing, even more genes were detected to be highly differentially regulated.

The dynamic nature of gene expression detected with cDNA sequencing was indicated by the larger number of genes to which highly different rank numbers were assigned for the two tested conditions. The threshold for rank numbers representative for low abundance genes (RPKM ≤10) was approximately 3800 and 3200 for the 10°C and 30°C samples respectively. The shift in the threshold of rank numbers indicated furthermore that more genes were up-regulated in response to cold shock than down-regulated, as fewer genes were detected with low RPKM at 10°C. These findings regarding the rank numbers of signal intensities were confirmed by determination of the numbers of genes significantly expressed and differentially regulated by more than two-fold (Table 3.15). Gene expression was considered to be significant if $RPKM_{30°C} > RPKM_{10°C} + 3\sqrt{RPKM_{10°C}}$ (or vice versa) for Illumina cDNA sequencing or if p-value ≤ 0.05 and FDR ≤ 0.05 could be calculated for microarray data.

Table 3.15 Total number of genes significantly expressed in response to cold shock.

Number of genes	Illumina	Affymetrix	Progenika
*Significantly differentially expressed	2933	2843	124
Differentially regulated, fold-change ≥ 2.0	2337	1281	122
Up-regulated	1478 (63.2%)	779 (60.9%)	61 (50%)
Down-regulated	859 (36.8%)	502 (39.1%)	61 (50%)

* Genes that showed significant differential expression according to p-value ≤ 0.05 and FDR ≤ 0.05 independent of absolute fold-change.

Strikingly, the total number of significantly differentially expressed genes derived from Progenika microarrays was significantly lower, whereas the number of genes detected with either Illumina or Affymetrix was comparable. However, considering only the differentially expressed genes, nearly all genes detected with Illumina were also differentially regulated by more than two-fold (85%). This result is in contrast to Affymetrix, where less than 50% of significantly expressed genes showed fold-changes higher than two. The ratio of up and down-regulated genes was also similar when comparing Illumina and Affymetrix. The overrepresentation of up-regulated genes was already indicated by the threshold of rank numbers representative for low abundance genes at 30°C and 10°C (see above). The Progenika microarrays were somehow limited in detecting this as here the same amount of genes was detected to be either up or down-regulated.

The rank number evaluation in combination with the examination of the total number of significantly expressed genes reflected important features.

- The gene expression profile derived from cDNA sequencing was highly dynamic displayed by:
1) The strong separation when comparing rank numbers of RPKM values (Fig. 3.21, A),
2) The high absolute fold-change with many genes found to be differentially regulated more than 50-fold (Fig. 3.21, B) and
3) The majority of sufficiently expressed genes were also significantly differentially regulated.

- Transcriptome analysis based on cDNA sequencing facilitated the detection of minor differences in gene regulation, since a larger range in absolute fold-change allowed a more precise graduation. This is illustrated by the comparison of Illumina and Affymetrix where the total number of genes sufficiently expressed was comparable but differed significantly when considering a significantly differential regulation.

- Evaluation of signal intensities in the case of microarrays was limited in reflecting the absolute fold-change, especially for the Progenika microarrays where few genes were found to be differentially regulated more than 10-fold.
- The Progenika microarrays were overall limited in detecting differentially expressed genes with statistical significance. The total number of significantly differentially regulated genes that were detected with the two microarrays differed by about one order of magnitude.

Considering only the total number of significantly expressed genes, Illumina and Affymetrix seemed to be broadly comparable. However, an intersection of the significantly differentially expressed genes showed that the gene lists did not completely overlap. For example, about 15% of the genes detected with Affymetrix could not be verified by the cDNA sequencing as shown in Figure 3.22.

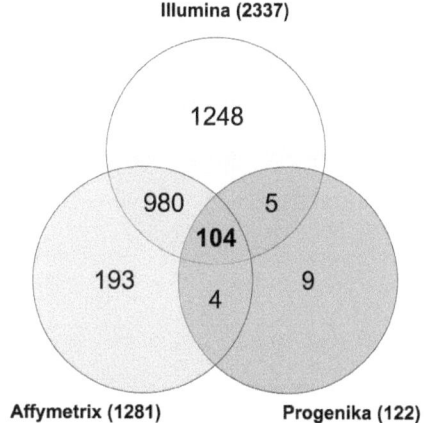

Fig. 3.22 Venn-diagram of significantly expressed genes displaying a differential regulation higher than two-fold that were detected with the three transcriptome platforms. Genes were considered to be significantly differentially expressed if calculated p-value and FDR were ≤ 0.05 and absolute fold-change ≥ 2.0 for genome microarrays and if $RPKM_{10°C} > RPKM_{30°C} + 3\sqrt{(RPKM_{30°C})}$ (or vice versa) for Illumina cDNA sequencing.

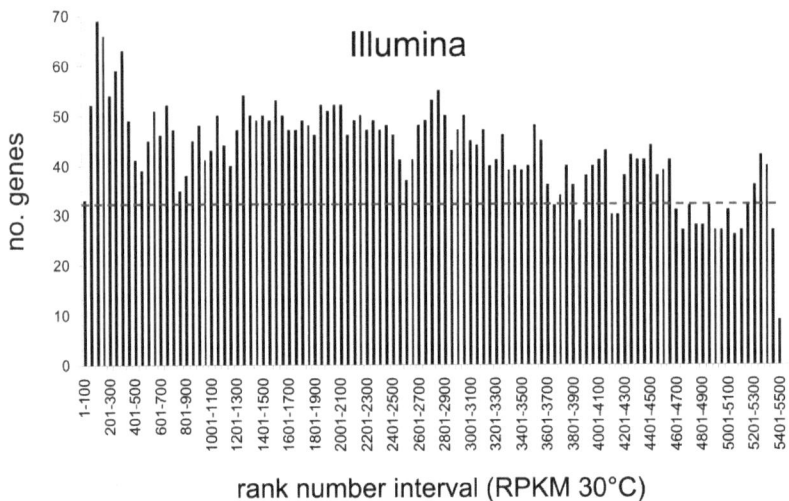

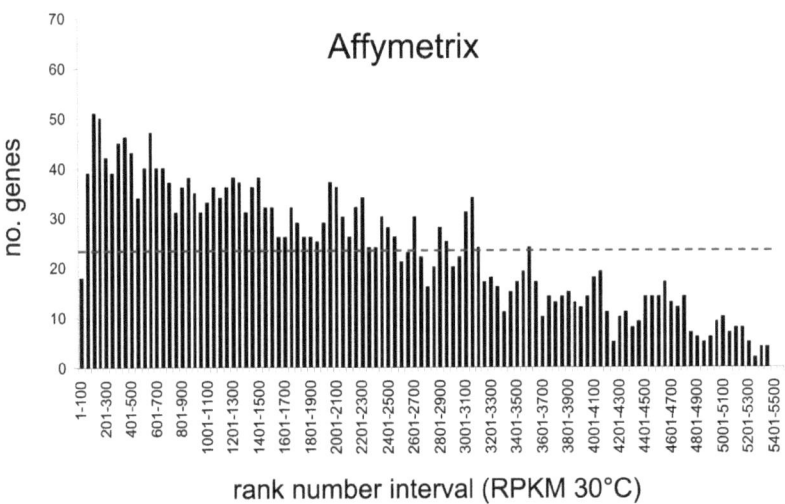

Fig. 3.23 Intersection of significant regulated genes obtained from either the Affymetrix or the Illumina results into rank number intervals. The numbers of significant differentially expressed genes detected with either Affymetrix microarrays or Illumina deep RNA sequencing were assigned to Illumina rank number intervals of 100 in steps of 50 (1-100, 51-150, 101-200, etc.). In total 1282 genes were detected with the Affymetrix microarray significantly expressed more than two-fold and 2337 genes with the Illumina deep RNA sequencing. Dashed lines indicate average number of genes per interval (Illumina 32, Affymetrix 23).

Interestingly, though the number of genes detected with the Progenika microarrays was very low, most of the genes (85%) identified by this analysis intersected with both the Illumina and Affymetrix results. This was an important finding regarding the transcriptome analysis with Progenika microarrays. Though the overall detection rate was rather poor with only 122 genes found to be significantly differentially expressed, implicating that less than 3% of all genes were differentially regulated in response to cold shock, the resulting gene list is reliable and the identified genes play an important role in response to cold shock as verified by the two other transcriptome platforms. From this, one can conclude that single regulated genes from certain pathways are representative for the whole pathway involved in the response, though not every single gene thereof was identified.

The advantages of transcriptome analysis based on cDNA sequencing in contrast to microarray analysis were already emphasized in Figure 3.21. An additional analysis however highlighted and strengthened this view even more. For this, significantly differentially expressed genes (for Illumina 2337 and for Affymetrix 1281) were assigned in rank number intervals according to RPKM values of the standard condition. Rank number intervals were allocated in windows of 100 in steps of size 50 (50 rank number overlap: 1-100, 51-150, 101-200 etc.), and the number of significantly differentially expressed genes falling into the respective intercepts were counted (see Figure 3.23). This counting was done separately for the Illumina and Affymetrix results, but in both cases assigned to rank number intervals of Illumina RPKM values.

The mean number of genes per intercept was 43 for Illumina and 23 for Affymetrix. Regarding the two graphs representative for Illumina and Affymetrix, the overall distribution showed striking differences. The significantly regulated genes detected with cDNA sequencing were constantly distributed over all intercepts. In contrast, the distribution of genes identified by Affymetrix microarrays showed a linear decrease corresponding to increasing rank numbers. As high rank numbers indicate a low RPKM value or signal intensity representative for low abundant transcript, this analysis demonstrated the limitation of microarrays in detecting lowly but sufficiently expressed genes, as from rank number 3100 on the significantly differentially regulated genes identified by Affymetrix was below average. In contrast, even low abundance transcripts at 30°C assigned with high rank numbers were identified by Illumina to be significantly regulated.

Discussion

DNA microarray platforms have been the technology of choice for transcriptome analysis during the last decade, however deep RNA sequencing is an emerging field in gene expression profiling which has already been applied to a few bacterial transcriptomes (Sittka *et al.*, 2009; Jäger *et al.*, 2009; Oliver *et al.*, 2009; Sharma *et al.*, 2010; Filiatraut *et al.*, 2010). This new technology presents a promising alternative to microarrays since it enables additionally the detection of low abundant transcripts, identification of transcription start sites and novel transcripts encoding for as yet unidentified proteins or sRNAs. Beside these advantages cDNA sequencing furthermore overcomes the major limitations of microarrays that are technically biased due to background levels of hybridization, saturation of signal intensities resulting in constricted detection of absolute fold-changes and are furthermore heavily dependent on the chosen probe-sets (Gautier *et al.*, 2004; Marioni *et al.*, 2008).

However, deep cDNA sequencing is not completely free of technical bias. Illumina sequencing comprises adapter ligation to cDNA-fragments and is still PCR based. This might lead to amplification artefacts due to incorporation of wrong nucleotides in case of repetitive nucleotides or extreme GC-content. A study by Kozarewa *et al.* (2009), albeit of genomes not transcriptomes, showed that amplification-free generation of the libraries improved the outcome when mapping reads to GC-biased genomes. A bias was observed when AT-rich genomes were examined. The analysis of *Plasmodium falciparum*, which has a GC-content of 22% still showed a significant shift away from the theoretical value towards a higher GC-content in the reads. This was not the case for results of *Bordetella pertussis* with a high genomic GC-content of ~ 68% that is in a comparable range to that of *P. putida* (61.5%).

Though deep RNA sequencing is not free from technical biases, recent studies comparing deep RNA sequencing and DNA microarrays suggest that beside the above mentioned advantages sequencing-based expression analyses is more robust and comparable and thus might displace microarray hybridization as the technology of choice for gene expression profiling in the future ('t Hoen *et al.*, 2008; Marioni *et al.*, 2008).

The comparison of the two microarray platforms showed major differences in the detection of significantly regulated genes. With Progenika microarrays merely 122 genes could be identified in contrast to the 1281 gene detected with Affymetrix microarrays which is a difference of one order of magnitude. Though the overlap of genes detected with both microarrays was high (108 out of 122 genes detected with Progenika intersected with the Affymetrix result), the vastly different sensitivity of the two microarrays makes a direct comparison difficult.

A comparative study of hippocampal gene expression of wild type and transgenic mice (δC-doublecortin-like kinase) analyzed with five different microarray platforms showed that minor differences in gene expression were more likely detected by a combination of different microarrays than by increasing the number of biological and technical replicates. Hence using different microarray platforms might provide complementary insights into biological processes (Pedotti *et al.*, 2008). The number of identified genes differed significantly from four genes identified by the respective Applied Biosystems microarray to 3051 genes by Agilent microarrays with a fixed FDR of ≤ 0.1. The four genes identified by the Applied Biosystems platform were found to be significantly differentially expressed by at least three other platforms.

Though the difference of the biological background of the hippocampal gene expression study was very distinct, these results supported the outcome of the microarray comparison in this work that results from different microarray platforms are rather difficult to compare directly.

As most of the genes detected by Progenika were verified by Affymetrix, the outcome of the Progenika analysis might be seen as basic cold stress response and Affymetrix data might be complementary to find genes that are also involved in the regulatory pathways identified by Progenika. The biological context of cold stress response however is discussed in the next chapter.

The same samples used for the comparative study of the five different microarray platforms (Pedotti *et al.*, 2008) were also sequenced using the Illumina whole genome sequencing ('t Hoen *et al.*, 2008). The major findings were consistent with results from this study. Though other microarrays were used and the biological context was diverse, the microarray designed by Affymetrix showed the best correlation with Illumina sequencing results with a Pearson correlation factor of 0.63. The number of significantly regulated genes was higher with 3179 (FDR 8.5%) than found with microarrays. Furthermore, the absolute fold-changes were larger and biological process with minor differences in the two samples could be detected with Illumina deep RNA sequencing but were missed with the microarrays.

In contrast to the outcome of this study, Bloom *et al.* (2009) saw major limitations in detecting low abundant transcripts by cDNA sequencing. By comparison of gene expression profiles of *Saccharomyces cerevisiae* derived from Illumina cDNA sequencing and Agilent two-channel microarrays, they observed that differential gene expression detected only by microarrays had significantly lower read coverage and vice versa, ORFs with read coverage above average were more likely to be detected only by sequencing.

In this study however, precisely cDNA sequencing revealed a high number of low abundant transcripts that were significantly differentially expressed. Out of the 86 identified low abundance transcripts (see Table 3.6) only four genes were also detected with the Affymetrix microarrays. One reason for the discrepancy might be the different definition of significantly differential expression. Bloom and co-workers used a Fisher's exact test with additional criteria of FDR ≤ 0.1 and fold-change ≥ 1.5. In this work genes were considered to be significantly differentially expressed if $RPKM_{30°C} > RPKM_{10°C} * 3\sqrt{RPKM_{10°C}}$ or vice versa and fold-change ≥ 2.0. Furthermore they mentioned the bias due to a high number of reads that could still be mapped to rRNA genes, though they enriched their samples for mRNA. We applied total RNA to cDNA synthesis and consequently to Illumina cDNA sequencing. Though about 91% of all reads were mapped to rRNA genes, still the approximately 300,000 non-rRNA reads per sample provided a sufficient coverage for transcriptome analysis and comparison of transcriptome platforms. Bloom *et al.* addressed the limited mRNA coverage by combining reads from several parallel Illumina sequencing runs for the same cDNA sample, at the price of increasing costs. The read coverage problem is becoming less important as future developments of sequencing technologies are already leading to an increasing number of reads per run. In the mid-term both are available. Costs for sequencing have already declined over the past few years enabling the usage of parallel lanes at the same costs as a few years ago, and as the read length has already been extended from 36 to 75 bp the coverage is duplicated by the same number of reads. Furthermore, with the emergence of other vendors such as Pacific Biosciences, Applied Biosystems competition will increase and thus drive down costs.

In summary, though deep cDNA sequencing is still subject to technical bias, it showed major advantages in gene expression analysis over the microarrays. Firstly, it showed the real abundance of transcripts since it is an open ended counting technique which is not affected by signal intensity saturation. Secondly, genes not detected by Affymetrix due to hybridization problems of AT-rich fragments could be identified by sequencing. Furthermore, the analysis of genes not detected with Affymetrix but with sufficient expression based on the sequencing results emphasized the problem that only predefined sequences can be detected due to the microarray design. As already mentioned in more detail in the previous chapter, cDNA sequencing revealed a high number of low abundant transcripts with significantly differential expression and transcripts in intergenic regions. These transcripts were not detected with microarrays due to low signal intensities or they were missed as the corresponding genomic loci were not represented on the microarray.

The comparison of the three platforms showed that the Progenika microarrays are less sensitive than the Affymetrix microarrays by detecting less significantly regulated genes in a range of one order of magnitude. This could be mainly ascribed to the design of the Progenika microarrays. However, the intersection of the three transcriptome platforms exhibited that the outcome is still reliable since the genes identified by Progenika microarrays were also detected with the two other platforms.

As previously mentioned, this project was incorporated into the PSYSMO consortium (http://www.psysmo.org/), one of eleven projects working on "Systems Biology of Microorganisms (SysMO). Beside the standardization of growth conditions (see 3.2), the consortium decided furthermore to use the Progenika microarrays for transcriptome analysis. The main reason for this was that at that time point the Progenika microarrays were the only genome microarrays directly available. The development of a new *P. putida* genome microarray by Affymetrix or an alternative company such as Agilent would have lasted too long for the design to start with transcriptome analysis within the next few months and would have been too costly in comparison to the Progenika microarrays. The Affymetrix microarrays used in this study were leftovers from a previous study and thus were limited to the transcriptome analysis of the wild type. For this reason, the Progenika microarrays were used for the transcriptome analysis of cold sensitive mutants (see chapter 3.8).

3.7 Key Players in Cold Shock Stress of *Pseudomonas putida* KT2440

The soil bacterium *P. putida* experiences temperature shifts in its natural habitat. Here, a combinatorial transcriptome analysis including three transcriptome platforms has been conducted to reveal the key players in cold shock response that allow the bacterium to survive after and adapt to a temperature down-shift from 30°C to 10°C.

The comparability of the three platforms has been discussed in detail in the previous chapter (3.6). For the biological interpretation of induced and repressed genes in response to cold shock, only those were considered that were detected by all three platforms and hence were considered to be the key players representative for the affected biological processes. If not otherwise mentioned, expression data such as fold-change correlate to results obtained from Illumina cDNA sequencing.

Furthermore, samples derived from the same cultures used for transcriptome analysis were additionally applied to metabolome and proteome analysis as an integrative approach to see if identified genes that were found to be important in cold shock response on the transcriptional level could be reflected in the metabolite profile and protein spectrum after cold shock.

3.7.1 Transcriptome analysis

Genes were considered to be significantly differentially expressed if p-value and FDR ≤ 0.05 and absolute fold-change ≥ 2.0 for microarrays and the normalized reads $RPKM_{30°C} > RPKM_{10°C} + 3\sqrt{RPKM_{10°C}}$ (or vice versa) for Illumina cDNA sequencing. In total, 2337, 1281 and 122 genes for Illumina, Affymetrix and Progenika respectively were found to be significantly expressed and differentially regulated by more than two-fold. Of these genes, 104 were found with all three platforms whereas one, PP1624, encoding a group II intron-encoding maturase) differed in the direction of regulation. This gene was found to be up-regulated with the microarrays but down-regulated with cDNA sequencing, and thus was excluded from further analysis. Since the sensitivity of the Progenika microarrays was considered to be low with one order of magnitude less detected genes, the defined threshold for genes to be significantly differentially expressed was changed to FDR ≤ 0.1 for Progenika microarray results. This increased the number of regulated genes to 175 whereof 159 had been found with the two other platforms as well, and thus were still considered to be significant. Out of the 159 genes, 85 were repressed and 74 were induced in response to cold stress. A list of these 159 genes can be found in Table 3.16.

Table 3.16 Summary of genes repressed or induced in response to cold shock.

locus id	Progenika p-value	Progenika FDR	Progenika fold-change	Affymetrix p-value	Affymetrix FDR	Affymetrix fold-change	Illumina RPKM 30°C	Illumina RPKM 10°C	Illumina fold-change	encoded protein	gene name
DOWN-regulated genes											
* PP0083	0.000243	0.017776	-2.28	0.000000	<0.000001	-19.40	173.71	2.42	-71.83	tryptophan synthase, beta subunit	trpB
* PP0268	0.000056	0.010971	-2.88	0.000000	<0.000001	-4.43	853.01	109.33	-7.80	outer membrane protein OprE3	oprQ
PP0282	0.003378	0.093398	-2.22	0.000005	0.000033	-2.18	595.61	164.38	-3.62	amino acid ABC transporter, periplasmic amino acid-bi	
PP0364	0.002717	0.088500	-2.04	0.000000	<0.000001	-3.32	73.53	12.08	-6.09	biotin biosynthesis protein BioH	bioH
PP0545	0.003317	0.093361	-1.86	0.000000	0.000003	-5.50	176.80	13.55	-13.04	aldehyde dehydrogenase family protein	
* PP0704	0.000238	0.017760	-3.19	0.000000	<0.000001	-8.83	48.70	5.71	-8.52	RNA polymerase sigma-70 factor, ECF subfamily	
PP0817	0.003252	0.093361	-1.87	0.000000	<0.000001	-5.82	169.07	7.31	-23.13	aminotransferase, class I	
* PP0849	0.001025	0.046865	-2.41	0.000001	0.000005	-2.79	733.50	332.40	-2.21	nucleoside diphosphate kinase	ndk
PP0864	0.000025	0.006802	-4.39	0.000000	<0.000001	-7.39	286.01	21.12	-13.54	ornithine decarboxylase, putative	
PP0913	0.000067	0.010971	-3.34	0.000000	<0.000001	-6.53	740.33	81.81	-9.05	conserved hypothetical protein	
* PP0915	0.000010	0.005765	-3.92	0.000000	<0.000001	-4.57	991.99	153.08	-6.48	superoxide dismutase (Fe)	sodB
PP0920	0.003657	0.098720	-2.35	0.000000	<0.000001	-4.39	30.74	2.81	-10.96	xenobiotic reductase B	xenB
PP0989	0.001380	0.056668	-2.22	0.000000	<0.000001	-15.41	243.25	15.37	-15.83	glycine cleavage system H protein	gcvH-1
PP1068	0.003682	0.098805	-2.58	0.000000	<0.000001	-3.68	325.75	42.04	-7.75	amino acid ABC transporter, ATP-binding protein	
* PP1070	0.000388	0.024447	-2.22	0.000085	0.000350	-5.19	465.91	51.29	-9.08	amino acid ABC transporter, permease protein	
PP1071	0.002965	0.091760	-2.05	0.000001	0.000006	-2.04	1008.75	307.11	-3.28	amino acid ABC transporter, periplasmic amino acid-bi	
PP1084	0.000248	0.017944	-2.69	0.000000	<0.000001	-2.80	654.74	136.89	-4.78	antioxidant, AhpC/Tsa family	
PP1088	0.000045	0.010675	-3.99	0.000000	<0.000001	-4.36	1374.98	157.19	-8.75	argininosuccinate synthase	argG
PP1110	0.003065	0.092569	-2.21	0.000000	0.000004	-5.13	409.69	26.71	-15.34	serine O-acetyltransferase, putative	
PP1111	0.000202	0.016835	-3.96	0.000000	<0.000001	-7.48	341.04	59.76	-5.71	synthetase, putative	
PP1112	0.000530	0.029934	-2.35	0.000000	<0.000001	-7.54	162.74	14.58	-11.16	conserved hypothetical protein	
* PP1140	0.001048	0.046865	-2.86	0.000001	0.000005	-3.90	260.10	22.32	-11.65	branched-chain amino acid ABC transporter, permease p	braD
PP1188	0.002355	0.080267	-1.91	0.000000	0.000003	-2.90	921.55	262.70	-3.51	C4-dicarboxylate transport protein	dctA
PP1206	0.003257	0.093361	-2.13	0.000005	0.000034	-2.39	946.08	146.57	-6.45	porin D	oprD
* PP1249	0.000013	0.005781	-4.34	0.000000	0.000003	-4.69	150.08	15.41	-9.74	lipoprotein, putative	
PP1303	0.000080	0.011252	-2.76	0.000000	0.000003	-4.16	1223.06	147.64	-8.28	sulfate adenylyltransferase, subunit 2	cysD
* PP1304	0.001025	0.046865	-2.71	0.000000	<0.000001	-6.43	522.12	51.09	-10.22	sulfate adenylyltransferase, subunit 1/adenylylsulfat	cysNC
PP1346	0.000614	0.032600	-2.11	0.000000	<0.000001	-3.32	94.22	31.44	-3.00	glutamate N-acetyltransferase/amino-acid acetyltransf	argJ
PP1360	0.000011	0.005765	-5.86	0.000000	<0.000001	-22.04	3081.80	60.27	-51.13	chaperonin, 10 kDa	groES
PP1361	0.000008	0.005285	-3.53	0.000000	<0.000001	-27.36	3237.99	39.48	-82.01	chaperonin, 60 kDa	groEL
PP1491	0.002997	0.091760	-1.70	0.000002	0.000014	-2.43	60.92	9.10	-6.70	CheW domain protein	
* PP1592	0.000038	0.009288	-2.67	0.000001	0.000012	-3.42	1503.03	238.72	-6.30	translation elongation factor Ts	tsf
* PP1638	0.000053	0.010971	-3.16	0.000000	<0.000001	-10.41	303.58	18.89	-16.07	ferredoxin--NADP reductase	fpr

Table 3.16 continued

locus id	Progenika p-value	Progenika FDR	Progenika fold-change	Affymetrix p-value	Affymetrix FDR	Affymetrix fold-change	Illumina RPKM 30°C	Illumina RPKM 10°C	Illumina fold-change	encoded protein	gene name
PP1664	0.002028	0.073767	-2.20	0.000028	0.000138	-2.02	164.62	40.56	-4.06	phosphoribosylglycinamide formyltransferase	purN
PP1826	0.003015	0.091796	-1.71	0.000006	0.000038	-2.30	157.90	44.89	-3.52	hydrolase, isochorismatase family	
* PP1858	0.000025	0.006802	-3.43	< 0.000001	< 0.000001	-3.96	629.74	118.97	-5.29	translation elongation factor P	efp
PP1901	0.001952	0.072480	-1.80	0.000006	0.000035	-2.04	96.84	47.73	-2.03	conserved hypothetical protein	
PP1977	0.000073	0.011045	-2.41	0.000001	0.000010	-3.24	91.94	15.90	-5.78	glutamyl-tRNA synthetase	gltX
* PP1982	0.000006	0.004723	-4.82	< 0.000001	< 0.000001	-6.15	135.67	13.11	-10.35	heat shock protein IbpA	ibpA
PP1985	0.003326	0.093361	-2.24	< 0.000001	< 0.000001	-16.20	342.57	14.38	-23.83	3-isopropylmalate dehydratase, large subunit	leuC
* PP2149	0.000307	0.021312	-3.98	< 0.000001	< 0.000001	-3.97	977.24	130.76	-7.47	glyceraldehyde 3-phosphate dehydrogenase	gap-2
PP2265	0.003106	0.092569	-2.24	0.000001	0.000011	-2.41	62.94	20.68	-3.04	5,10-methylene-tetrahydrofolate dehydrogenase	folD-2
PP2299	0.001353	0.056393	-2.64	0.000002	0.000018	-3.04	996.08	216.70	-4.60	trigger factor	tig
PP2339	0.000202	0.016835	-3.58	< 0.000001	< 0.000001	-4.35	866.62	106.04	-8.17	aconitate hydratase 2	acnB
PP2432	0.002885	0.090895	-2.11	0.000014	0.000077	-3.84	38.41	4.51	-8.52	oxygen-insensitive NAD(P)H nitroreductase	
* PP2453	0.000065	0.010971	-3.93	< 0.000001	< 0.000001	-7.24	52.69	13.53	-3.90	L-asparaginase II	ansA
* PP2550	0.000428	0.026065	-3.21	< 0.000001	< 0.000001	-3.92	236.20	45.41	-5.20	conserved hypothetical protein TIGR01033	
PP2669	0.003661	0.098720	-2.17	< 0.000001	< 0.000001	-5.11	105.07	11.90	-8.83	outer membrane protein, putative	
PP2680	0.000435	0.026209	-5.44	< 0.000001	< 0.000001	-8.12	1065.53	65.83	-16.19	aldehyde dehydrogenase family protein	
* PP2817	0.000070	0.011045	-4.75	< 0.000001	< 0.000001	-18.70	138.79	5.18	-26.79	multidrug efflux RND membrane fusion protein MexC	mexC
* PP2818	0.000063	0.010971	-4.02	< 0.000001	< 0.000001	-21.58	143.19	6.59	-21.74	multidrug efflux RND transporter MexD	mexD
PP2819	0.000103	0.011252	-2.55	< 0.000001	< 0.000001	-12.38	99.39	6.28	-15.83	outer membrane protein OprJ	oprJ
* PP2903	0.000097	0.011252	-2.92	0.000001	0.000003	-3.56	644.98	88.29	-7.30	peptidyl-prolyl cis-trans isomerase B	ppiB
PP2928	0.002256	0.078877	-2.99	0.000001	0.000003	-3.30	1152.15	182.17	-6.32	conserved hypothetical protein	
PP3155	0.000213	0.017451	-3.44	< 0.000001	< 0.000001	-8.18	178.40	16.59	-10.75	outer membrane ferric siderophore receptor, putative	
* PP3612	0.000092	0.011252	-3.97	< 0.000001	< 0.000001	-14.48	162.73	8.91	-18.26	TonB-dependent receptor, putative	
PP4127	0.003162	0.092623	-2.25	< 0.000001	< 0.000001	-4.05	315.32	76.52	-4.12	NADH dehydrogenase I, J subunit	nuoJ
PP4131	0.002920	0.091476	-2.60	0.000003	0.000003	-3.33	200.02	60.10	-3.33	NADH dehydrogenase I, N subunit	nuoN
PP4178	0.000225	0.017760	-2.72	< 0.000001	< 0.000001	-7.23	88.96	12.18	-7.30	dienelactone hydrolase family protein	
PP4186	0.000018	0.006030	-6.99	< 0.000001	< 0.000001	-7.79	10061.23	578.02	-17.41	succinyl-CoA synthetase, beta subunit	sucC
PP4187	0.000184	0.016518	-5.89	< 0.000001	0.000181	-7.38	2333.07	168.06	-13.88	2-oxoglutarate dehydrogenase, lipoamide dehydrogenase	lpdG
PP4191	0.001596	0.063589	-2.14	0.000039	0.000181	-2.13	1983.70	556.40	-3.57	succinate dehydrogenase, flavoprotein subunit	sdhA
PP4244	0.000202	0.016835	-2.85	< 0.000001	< 0.000001	-12.66	108.17	16.66	-6.49	ferric regulator Pfrl	pfrl
PP4256	0.000809	0.040979	-2.40	< 0.000001	< 0.000001	-4.23	200.38	43.57	-4.60	cytochrome c oxidase, cbb3-type, subunit II	ccoO-2
PP4385	0.000333	0.021928	-2.53	< 0.000001	< 0.000001	-3.70	86.73	29.99	-2.89	flagellar basal-body rod protein FlgG	flgG
PP4678	0.000015	0.005854	-6.79	< 0.000001	< 0.000001	-12.56	3671.37	95.59	-38.41	ketol-acid reductoisomerase	ilvC
PP4679	0.000016	0.005854	-5.28	< 0.000001	< 0.000001	-9.24	2466.64	197.81	-12.47	acetolactate synthase, small subunit	ilvN
PP4680	0.002858	0.090581	-2.88	< 0.000001	< 0.000001	-5.74	2192.74	174.13	-12.59	acetolactate synthase 3 catalytic subunit	ilvB

Table 3.16   continued

locus id	Progenika			Affymetrix			Illumina			encoded protein	gene name
	p-value	FDR	fold-change	p-value	FDR	fold-change	RPKM 30°C	RPKM 10°C	fold-change		
* PP4727	0.000933	0.045165	-4.82	<0.000001	<0.000001	-11.97	1591.50	39.75	-40.04	dnaK protein	dnaK
PP4728	0.002517	0.083167	-3.34	<0.000001	<0.000001	-32.27	932.79	47.55	-19.62	heat shock protein GrpE	grpE
* PP4870	0.000074	0.011045	-4.27	<0.000001	<0.000001	-9.27	782.08	117.99	-6.63	azurin	
PP4960	0.000258	0.018371	-2.72	<0.000001	<0.000001	-5.32	491.68	105.11	-4.68	fructose-1,6-bisphosphate aldolase	fda
PP4961	0.001787	0.069194	-2.09	<0.000001	0.000003	-3.20	294.06	35.78	-8.22	lipoprotein, putative	
PP4965	0.000748	0.038967	-2.14	0.000089	0.000359	-2.11	342.71	94.32	-3.63	transketolase	tktA
* PP5046	0.000791	0.040451	-4.28	<0.000001	<0.000001	-14.14	2128.00	66.98	-31.77	glutamine synthetase, type I	glnA
PP5075	0.001537	0.062184	-2.08	<0.000001	<0.000001	-5.07	353.77	58.11	-6.09	glutamate synthase, small subunit	gltD
PP5076	0.001895	0.070834	-2.77	<0.000001	<0.000001	-4.04	447.40	74.82	-5.98	glutamate synthase, large subunit	gltB
PP5128	0.001680	0.065598	-2.20	0.000004	0.000029	-2.73	426.25	89.52	-4.76	dihydroxy-acid dehydratase	ilvD
PP5170	0.003331	0.093361	-1.79	<0.000001	<0.000001	-4.82	78.85	7.20	-10.96	sulfate ABC transporter, permease protein	cysT
* PP5171	0.000165	0.015651	-2.54	<0.000001	<0.000001	-9.73	170.79	5.73	-29.83	sulfate ABC transporter, periplasmic sulfate-binding	cysP
PP5215	0.002322	0.080267	-3.22	<0.000001	0.000004	-3.28	304.95	44.73	-6.82	thioredoxin	trx-2
PP5232	0.000002	0.004607	-7.02	<0.000001	<0.000001	-18.03	1230.70	8.95	-137.58	conserved hypothetical protein	
* PP5412	0.000055	0.010971	-3.02	<0.000001	<0.000001	-4.94	1351.18	196.67	-6.87	ATP synthase F1, epsilon subunit	atpC
* PP5413	0.000098	0.011252	-3.51	<0.000001	<0.000001	-4.12	3992.04	506.91	-7.88	ATP synthase F1, beta subunit	atpD
* PP5414	0.000063	0.010971	-3.36	<0.000001	<0.000001	-3.45	5645.62	855.55	-6.60	ATP synthase F1, gamma subunit	atpG

UP-regulated genes

locus id	p-value	FDR	fold-change	p-value	FDR	fold-change	RPKM 30°C	RPKM 10°C	fold-change	encoded protein	gene name
* PP0021	0.000405	0.024926	2.42	<0.000001	<0.000001	7.92	27.17	318.86	11.74	hypothetical protein	
PP0090	0.000479	0.028214	2.51	<0.000001	0.000005	3.45	12.11	49.71	4.11	conserved hypothetical protein	
* PP0153	0.000021	0.006600	6.34	<0.000001	<0.000001	7.34	66.66	205.33	3.08	conserved hypothetical protein	
PP0185	0.000313	0.021447	2.91	<0.000001	<0.000001	21.30	4.82	483.25	100.26	alginate biosynthesis protein PprA	pprA
PP0330	0.000365	0.023255	2.78	<0.000001	<0.000001	15.24	13.54	222.40	16.43	conserved hypothetical protein	
PP0386	0.001333	0.056393	3.39	<0.000001	<0.000001	6.03	5.74	36.96	6.44	sensory box protein	
PP0584	0.001832	0.070667	2.08	<0.000001	0.000001	3.79	22.13	101.49	4.59	methyl-accepting chemotaxis transducer	
PP0624	0.001865	0.070667	2.09	0.000002	0.000014	4.35	24.21	139.20	5.75	conserved hypothetical protein TIGR00726	
PP0739	0.000137	0.013771	2.71	<0.000001	<0.000001	8.17	52.18	446.98	8.57	deoxyribodipyrimidine photolyase	phrB
* PP0741	0.000091	0.011252	2.72	<0.000001	<0.000001	8.79	51.82	425.65	8.21	conserved hypothetical protein	
PP0742	0.000101	0.011252	3.17	<0.000001	<0.000001	14.20	21.80	235.81	10.82	conserved hypothetical protein	
PP0743	0.002435	0.081776	1.70	<0.000001	<0.000001	5.86	23.76	172.36	7.25	conserved hypothetical protein	
PP0810	0.002758	0.088500	1.99	<0.000001	0.000001	6.49	130.85	537.36	4.11	cyoups1 protein	cyoups1
* PP0831	0.001043	0.046865	2.46	<0.000001	<0.000001	12.97	19.61	225.46	11.50	hypothetical protein	

Table 3.16 continued

locus id	Progenika p-value	Progenika FDR	Affymetrix fold-change	Affymetrix p-value	Affymetrix FDR	RPKM 30°C	Illumina RPKM 10°C	Illumina fold-change	encoded protein	gene name
* PP0868	0.000319	0.021578	3.15	<0.000001	<0.000001	11.79	27.87	363.75	13.05	glycine betaine/carnitine/choline ABC transporter	
* PP0870	0.000234	0.017760	2.68	<0.000001	<0.000001	16.80	26.66	547.34	20.53	glycine betaine/carnitine/choline ABC transporter	
* PP0871	0.000064	0.010971	3.33	<0.000001	<0.000001	12.19	53.05	510.98	9.63	glycine betaine/carnitine/choline ABC transporter	
PP1023	0.000012	0.005765	3.65	<0.000001	<0.000001	6.16	68.91	404.27	5.87	6-phosphogluconolactonase	pgl
* PP1075	0.003268	0.093361	2.29	<0.000001	<0.000001	6.77	26.29	147.25	5.60	glycerol kinase	glpK
PP1082	0.002135	0.076130	2.42	<0.000001	<0.000001	4.54	45.45	93.33	2.05	bacterioferritin	bfr
* PP1099	0.000402	0.024926	3.07	<0.000001	<0.000001	3.76	805.79	4013.32	4.98	cold shock domain family protein	
PP1147	0.000335	0.021928	2.27	<0.000001	<0.000001	6.98	10.89	89.46	8.21	hypothetical protein	
* PP1148	0.000028	0.007149	3.08	<0.000001	<0.000001	7.37	18.47	288.24	15.61	hypothetical protein	
* PP1186	0.000170	0.015840	2.68	<0.000003	<0.000001	3.93	306.98	1082.47	3.53	transcriptional regulatory protein PhoP	phoP
PP1187	0.000005	0.004723	3.53	<0.000001	<0.000001	6.95	93.17	673.46	7.23	sensor protein PhoQ	phoQ
PP1252	0.003505	0.095937	2.25	<0.000010	<0.000001	2.87	22.69	57.99	2.56	group II intron-encoding maturase	
* PP1428	0.001038	0.046865	2.63	<0.000001	<0.000001	7.77	279.37	2544.04	9.11	sigma factor algU negative regulatory protein MucA	mucA
* PP1429	0.000014	0.005854	3.81	<0.000001	<0.000001	9.33	144.81	1171.14	8.09	sigma factor algU regulatory protein AlgN	algN
PP1451	0.000202	0.016835	3.25	<0.000050	<0.000223	2.58	25.87	74.37	2.87	conserved hypothetical protein	
PP1515	0.001339	0.056393	1.93	<0.000001	<0.000001	6.22	5.59	82.65	14.78	transcriptional regulator, tetR family	
PP1624	0.000239	0.017760	2.40	<0.000007	<0.000045	3.05	35.30	12.43	-2.84	group II intron-encoding maturase	
* PP1689	0.000088	0.011252	3.85	<0.000001	<0.000001	15.33	110.47	2179.99	19.73	long-chain fatty acid transporter, putative	
PP1690	0.000133	0.013611	3.91	<0.000001	<0.000001	279.11	7.43	3614.76	486.51	conserved hypothetical protein	
* PP1691	0.000006	0.004723	11.14	<0.000001	<0.000001	165.65	16.00	19708.26	1231.77	conserved hypothetical protein	
* PP1797	0.000159	0.015379	2.64	<0.000001	<0.000001	10.52	24.34	377.61	15.51	HlyD family secretion protein	
PP1846	0.001682	0.065598	2.45	<0.000014	<0.000074	4.14	12.61	72.49	5.75	ATP-dependent RNA helicase, DEAD box family	
PP1868	0.000232	0.017760	4.55	<0.000001	<0.000001	5.44	294.51	1326.94	4.51	conserved hypothetical protein	
* PP1910	0.000118	0.012258	2.53	<0.000001	<0.000004	2.62	1047.08	3361.95	3.21	conserved hypothetical protein	
PP1912	0.000853	0.042415	2.04	<0.000001	<0.000003	5.37	301.56	1363.73	4.52	fatty acid/phospholipid synthesis protein PlsX	plsX
PP2166	0.000984	0.046381	1.97	<0.000007	<0.000041	3.94	66.91	183.18	2.74	anti-anti-sigma factor	
* PP2240	0.000494	0.028691	2.39	<0.000001	<0.000007	6.08	7.62	123.67	16.23	ABC efflux transporter, permease/ATP-binding protein	
* PP2296	0.000000	0.001084	9.80	<0.000001	<0.000001	30.08	80.52	1488.08	18.48	hypothetical protein	
* PP2297	0.000276	0.019428	4.29	<0.000001	<0.000001	43.34	21.74	953.65	43.87	integrative genetic element Ppu40, integrase	
PP2512	0.000104	0.011252	3.72	<0.000001	<0.000001	13.06	12.80	399.42	31.20	GTP cyclohydrolase I	folE-2
PP2730	0.003373	0.093398	2.54	<0.000001	<0.000001	23.01	6.69	120.80	18.06	lipoprotein, putative	
* PP2732	0.001142	0.049895	2.60	<0.000001	<0.000001	6.40	2.35	61.72	26.26	conserved hypothetical protein	
PP2735	0.000498	0.028691	2.70	<0.000001	<0.000001	7.37	4.43	65.48	14.78	conserved hypothetical protein	
PP2737	0.002750	0.088500	3.27	<0.000001	<0.000001	7.14	4.76	105.67	22.20	oxidoreductase, short-chain dehydrogenase/reductase	

Table 3.16 continued

locus id	Progenika p-value	Progenika FDR	Progenika fold-change	Affymetrix p-value	Affymetrix FDR	Affymetrix fold-change	Illumina RPKM 30°C	Illumina RPKM 10°C	Illumina fold-change	encoded protein	gene name
* PP2738	0.000883	0.043508	3.67	<0.000001	<0.000001	11.87	8.49	111.59	13.14	transcriptional regulator, putative	
PP2883	0.001077	0.047440	2.10	<0.000001	<0.000001	4.43	11.98	78.75	6.57	hypothetical protein	
* PP2949	0.000530	0.029934	3.09	0.000011	0.000063	2.82	8.28	27.19	3.28	membrane protein, putative	
PP3039	0.001227	0.052768	3.93	<0.000001	<0.000001	6.35	9.90	97.56	9.85	pyocin R2_PP, conserved hypothetical protein	
* PP3099	0.000915	0.044683	2.06	<0.000001	<0.000001	4.91	90.65	393.85	4.34	conserved hypothetical protein	
PP3316	0.000088	0.011252	5.15	<0.000001	<0.000001	8.90	2.54	126.19	49.68	chaperone-associated ATPase, putative	
* PP3317	0.000441	0.026236	4.47	<0.000001	<0.000001	25.20	28.67	329.63	11.50	hypothetical protein	
PP3379	0.003264	0.093361	1.92	0.000001	0.000004	2.68	4.60	11.33	2.46	epimerase KguE, putative	
* PP3433	0.000102	0.011252	10.02	<0.000001	<0.000001	27.46	33.30	730.30	21.93	4-hydroxyphenylpyruvate dioxygenase	hpd
PP3434	0.001011	0.046865	9.19	<0.000001	<0.000001	36.50	33.49	220.04	6.57	hypothetical protein	
PP3443	0.001451	0.059115	3.00	<0.000001	<0.000001	3.58	13.28	67.26	5.06	glyceraldehyde-3-phosphate dehydrogenase, putative	
PP3563	0.002459	0.081776	1.79	0.000004	0.000030	2.14	4.21	27.67	6.57	conserved hypothetical protein	
* PP3703	0.000564	0.030927	3.08	0.000005	0.000034	3.30	153.44	619.01	4.03	hypothetical protein	
* PP3704	0.000051	0.010971	4.53	<0.000001	<0.000001	7.48	116.66	355.91	3.05	hypothetical protein	
PP3765	0.001181	0.051213	2.77	<0.000001	<0.000001	5.97	39.92	139.37	3.49	transcriptional regulator MvaT, P16 subunit, putative	
* PP3767	0.000183	0.016518	2.52	<0.000001	<0.000001	7.37	74.03	802.66	10.84	conserved hypothetical protein	
PP3827	0.002952	0.091760	1.80	0.000001	0.000010	2.49	26.42	86.80	3.29	conserved hypothetical protein	
PP4020	0.001878	0.070667	2.33	0.000001	0.000009	3.56	5.56	31.97	5.75	oxidoreductase, putative	
PP4376	0.002681	0.088064	2.17	<0.000001	0.000003	3.90	994.75	3452.40	3.47	flagellar cap protein FliD	fliD
PP4377	0.002451	0.081776	2.71	<0.000001	<0.000001	3.97	1541.61	5004.98	3.25	flagellin FlaG, putative	
* PP4402	0.000227	0.017760	3.55	<0.000001	<0.000001	13.37	3.52	173.29	49.23	2-oxoisovalerate dehydrogenase, beta subunit	bkdA2
PP4547	0.001361	0.056393	2.62	<0.000001	<0.000001	12.93	2.63	129.46	49.23	glutamine synthetase, putative	
PP4548	0.000569	0.030927	3.07	<0.000001	<0.000001	23.39	8.22	96.83	11.78	oxidoreductase, putative	
PP4714	0.001809	0.069537	3.93	<0.000001	<0.000001	4.69	556.14	3775.73	6.79	conserved hypothetical protein	
* PP4813	0.000974	0.046322	2.26	<0.000001	0.000003	2.99	21.78	51.44	2.36	PAP2 family protein/DedA family protein	
* PP5203	0.000186	0.016518	2.51	<0.000001	0.000001	2.89	29.47	159.74	5.42	5-formyltetrahydrofolate cyclo-ligase family protein	

* Genes detected by Progenika microarrays displayed statistical significance of p-value ≤ 0.05 and FDR ≤ 0.05.

The genes with lower expression at 10°C could be grouped into five major functional groups: 1) amino acid transport and metabolism (24 genes), 2) energy production and conversion (14 genes), 3) cell wall biogenesis and maintenance (8 genes), 4) posttranslational modification, protein turnover and chaperones (8 genes), 5) transcription and translation (5 genes).

Generally, the spectrum of cold shock repressed genes indicated a reduction in major pathways of intermediate metabolism such as the citric acid cycle. Besides a general reduction in energy production and conversion as represented by a number of genes such as the NADH dehydrogenase (here *nuoJN*) or ATP synthase (*atpCDH*) cluster or representatives of the citric acid cycle such as *sucC*, *ipdG* and *sdhA*, protein biosynthesis in tandem with nucleotide and amino acid biosynthesis was mainly affected.

A graphical overview of down-regulated metabolic routes with the corresponding genes is given in Figure 3.24.

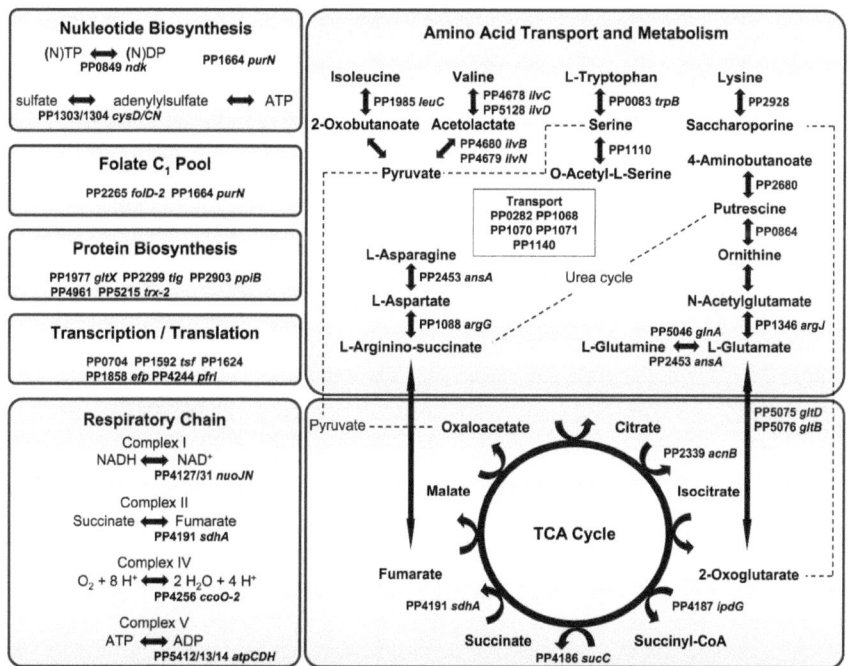

Fig. 3.24 Overview of down-regulated genes detected with all three transcriptome platforms. Genes that were identified to be significantly regulated fell mainly into functional categories of energy production and conversion, transcriptional and translational processes, *de novo* biosynthesis of either nucleotides or amino acids and one carbon pool by folate.

RESULTS AND DISCUSSION

In response to a rapid temperature down-shift the membrane fluidity of bacterial cells decreases and thereby affects membrane-associated functions such as transport and secretion. This was also reflected on the transcriptional level as many genes involved in cell wall biogenesis were found to be differentially regulated, such as TonB-dependent receptors (PP3612), (outer) membrane proteins (OprQ, OprD, PP2669, PP2949, PP3155) and putative lipoproteins (PP1249, PP2730, PP4961), as well as a number of transporter and secretion systems (PP1797).

The inner-membrane translocator *braD*, the last subunit of a branched-chain amino acid transporter, was down-regulated. The *braG/ilvG/ilvM/braD* transporter is associated with the translocation of leucine/isoleucine/valine (Hoshino *et al.*, 1992). This correlates very well with the identification of differentially expressed genes involved in the biosynthesis and degradation of valine, leucine and isoleucine. The first steps in the biosynthesis of valine from pyruvate via (S)-2-acetolactate catabolized by *ilvN/B* and 2-oxoisovalerate by *ilvC, ilvD* as intermediates, and of isoleucine via 2-oxobutanoate were down-regulated whereas the degradation of valine and isoleucine were increased as reflected by *bkdA-2*. This might reflect a shift from protein biosynthesis to the use of amino acids as precursors for the citric acid cycle as the degradation of valine and isoleucine leads to the production of acetyl-CoA and succinyl-CoA.

The RND efflux system MexCD/OprJ was also repressed in response to cold shock. It had already been shown to be important in antibiotic and solvent resistance (Poole, 2001). With regard to antibiotic or solvent resistance, an alteration of the outer membrane leads to a weaker permeability and in combination with an induction of efflux pumps to an export of causative agents at the same time. As the regulation of efflux pumps and transporter systems is an important mechanism in membrane alteration that is highly affected at low temperatures, it is likely that the induction or repression of MexCD/OprJ RND efflux system is a major regulatory mechanism of cell wall maintenance as indicated here by the high down-regulation (16 to 27-fold) of the efflux pump.

The last two subunits of the *cysAWTP* sulphate ABC transporter were also significantly down-regulated. Sulphate can serve as substrate for the synthesis of ATP when converted to adenylylsulphate and ApppppA (P1, P4-Bis (5'-adenosyl) tetraphosphate). Besides the sulphate transporter the sulphate adenylyltransferase subunits (*cysD/CN*) that convert sulphate to adenylylsulphate were also repressed supporting the specific reduction of ATP synthesis.

An ABC transporter (PP1068-1071), of which the first subunit was annotated to be involved in ectoine transport (http://www.pseudomonas.com) was down-regulated. The uptake of ectoine was already shown to be important as stress protectant against heat and high salinity in various halophilic species such as *Vibrio parahaemolyticus* (Naughton *et al.*, 2009), *Streptomyces coelicolor* (Bursy *et al.*, 2008), *Chromohalobacter salexigens* (García-Estepa *et al.*, 2006; Vargas *et al.*, 2008) and *Salinivibrio costicola* (Zhu *et al.*, 2008). A recent study on lipid monolayer and bilayer membranes by Harishchandra *et al.* (2010) demonstrated that the compatible solute ectoine might be important for repair mechanisms of the cell membrane in response to extreme temperatures or osmotic pressure by increasing its fluidity. Ectoine can be either synthesized from L-aspartate or serine (glycine). Based on the transcriptome data the synthesis of both amino acids seemed to be repressed indicating that the synthesis of ectoine might also be down-regulated.

The transcriptome data furthermore indicated that osmolytes such as glycine betaine and the amino acid glutamate were accumulated in response to cold shock as pathways affecting the conversion were down-regulated. The glycine cleavage system H protein GcvH-1 was significantly repressed by about 16-fold. Enzymes directly involved in the synthesis of glutamate were consistently repressed. *AnsA* (L-aspariginase II) and *glnA* (glutamine synthetase, type I) converting L-glutamine to L-glutamate or vice versa were down-regulated about four and 32-fold respectively, the glutamate synthase operon *gltD/B* converting 2-oxoglutarate to L-glutamate and *argJ*, a glutamate N-acetyltransferase six and three-fold respectively, and the composition of L-glutamyl-tRNA by *gltX* about six-fold.

Glycine betaine, a common and effective osmolyte, is dominant in salt-stressed cells, but also found as compatible solute in cells grown under low temperatures in different species such as *Listeria monocytogenes* (Ko *et al.*, 1994), *Yersinia enterocolitica* (Park *et al.*, 1995) and *Bacillus subtilis* (Brigulla *et al.*, 2003), suggesting a strong correlation of the response mechanisms to osmotic and cold stress. For *Listeria monocytogenes*, it was shown that the active uptake of glycine betaine was stimulated by low temperatures. This is supported by the transcriptome data here, as genes for the ABC transporter specific for glycine betaine were up-regulated 10 to 20-fold; three of the four subunit genes were detected with all three platforms.

In addition to the accumulation of glycine betaine and glutamate, the transcriptome data also indicated that the synthesis of glycerol that can also serve as an osmoprotectant was affected. The enzymatic reactions from glycerol to β-D-fructose 1,6-biphosphate were repressed. The aldehyde dehydrogenase family proteins PP0545 and PP2680, which convert D-glycerate to D-glyceraldehyde, the precursor of glycerol, were 13 and 16-fold down-regulated. D-glycerate can be converted to glycerate 3-phosphate, an intermediate in glycolysis/gluconeogenesis. The conversion from β-D-fructose 1,6-biphosphate to glycerate 3-phosphate was also repressed (*gap-2*, *fda*), whereas the direct conversion from glyceraldehyde 3-phosphate to glycerate 3-phosphate by PP3443 was induced. In contrast to the repression of the conversion of glycerol via glycolysis/gluconeogenesis, the conversion of glycerol to glycerophospholipids which are components of the cell membrane was induced (*plsX, glpK*).

Another gene was found to be highly affected by cold shock as indicated by the 16-fold down-regulation. PP2680 encoding an aldehyde dehydrogenase was described to be organized in a gene cluster responsible for the degradation 2-phenylethanol (Arias *et al.*, 2008). This gene cluster, referred to as the *ped* cluster in *P. putida* U, consists of 17 genes. The remaining 16 genes were not detected by the Progenika microarrays and showed only minor differential regulation according to the Affymetrix and Illumina results. As aldehyde dehydrogenases are involved in many processes by acting on aldehyde substrates, this gene cluster was not considered to be involved in cold shock response of *P. putida* wild type. This gene cluster will be discussed extensively in chapter 3.8 referring to the transcriptional profile of the cold sensitive mutants, and was therefore mentioned here.

The cold shock response was furthermore reflected by the repression of so-called heat shock genes (*groES, groEL, dnaK* and *grpE*). These heat shock genes are well characterized in *E. coli* and *B. subtilis* and are induced during growth above the optimal growth temperature (Bardwell & Craig, 1984; Holtmann *et al.*, 2004; Carruthers & Minion, 2009). Another transcriptome study by Phadtare & Inouye (2004) of the cold shock response in *E. coli* identified *groEL/ES* (*mopAB*) to be induced directly after cold shock suggesting that these chaperones play a global role in adaptation processes to different temperatures in *E. coli* and are not specified as heat shock proteins.

Here, *groEL* and *groES* were significantly down-regulated, according to cDNA sequencing by 82 and 51-fold, respectively. The molecular chaperones GroEL and GroES play an important role in protein folding and rapid protein degradation in case of dysfunctional proteins. The degradation by proteases is facilitated by building a complex with the target protein to be degraded. The rate-limiting step in this degradation is the trigger factor Tig. Studies in *E. coli* (Kandror *et al.*, 1995, 1997) showed that the trigger factor is required for the formation of the complex and regulates GroEL function as high levels of the trigger factor increased the binding affinity of GroEL to and degradation of the misfolded protein. In contrast to *E. coli* (Kandror & Goldberg, 1997) and *Pseudoalteromonas haloplanktis* (Piette *et al.*, 2010) where the trigger factor was identified as a cold-induced chaperone though associated with GroEL, it is repressed in *P. putida* (4.6-fold) suggesting a distinct role of the orthologs in different species.

Another heat shock protein complex belonging to the Hsp70 protein family was found to be repressed in response to cold shock, the genes *dnaK* and *grpE* were 40 and 20-fold down-regulated. *DnaJ* the third gene of this complex was not detected by the Progenika microarrays but found to be down-regulated with the two other platforms (7.5-fold with Affymetrix, 4.8-fold with Illumina). In conjunction with *dnaK* and *grpE*, *ibpA* was also repressed (10-fold). The encoded small chaperone binds denatured proteins and thereby facilitates the refolding by DnaJ/DnaK/GrpE (Veinger et al., 1998). In contrast to *Pseudomonas* syringae *pv. glycinea* (Keith et al., 1999) and *E. coli* (Ang & Georgopoulos, 1989) where DnaJ, GrpE and DnaK were required for growth under standard and increased temperatures, it seems that in *P. putida* the protein complex may be involved in other regulatory networks. A study by Dubern et al. (2005) in *P. putida* PCL1445 revealed that *dnaK* and *grpE* were not essential as transposon mutants affected in these genes were not impaired in growth under standard temperature, but confirmed their role as heat shock proteins. This suggests that there might be a compensatory heat shock mechanism of *dnaJ/dnaK/grpE* and *groEL/ES* in *P. putida*. In addition to their functional relevance as heat shock proteins, the genes were found to be relevant for the production of a lipopeptide, putisolvin that is involved in biofilm formation and degradation. This implicates that the genes might also be involved in cell wall biogenesis, an important mechanism in cold shock response.

The heat shock proteins DnaK/DnaJ/GrpE all have an AlgU specific promoter site (Firoved et al., 2002), thus the induction of *mucA* and *algN* correlated very well with the repression of the heat shock proteins since these are negative regulators of the sigma factor AlgU. Hence, an induction of MucA and AlgN that bind to AlgU leads to repression of *algU* and consequently low expression of the sigma factor-dependent promoter regions of *dnaK/dnaJ/GrpE*. Additionally the regulation of the alginate biosynthesis protein PprA, the homolog to AlgR in *P. aeruginosa*, is also AlgU dependent and the encoding gene was found to be strongly induced (100-fold) after cold shock. The strong correlation of alginate biosynthesis and cell wall maintenance has been displayed in many studies. For example Wood et al. (2006) demonstrated that alginate biosynthesis was activated by cell wall-inhibitory antibiotics and Govan & Deretic (1996) showed that alginate overproduction in *P. aeruginosa* CF isolates caused by *mucA* mutations is linked to mucoid phenotypes. In *P. aeruginosa* the AlgU-MucABCD operon is likely to form an inner membrane complex as MucA is located in the inner membrane (Mathee et al., 1997) and is thus likely also to be involved in the maintenance of the cell membrane in response to cold shock. Furthermore, MucA was described to be directly involved in the response to cell wall stress caused by either D-cycloserine (Wood & Ohman, 2009) or osmotic stress (Behrends et al., 2010). These data indicate that MucA and AlgN might be the major regulatory proteins in cold shock response.

Another transcriptional regulator, the two-component system PhoPQ, was also found to be induced. The PhoPQ two-component system is involved in virulence and polymyxin B resistance (Macfarlane et al., 2000; Gooderham et al., 2009; Schurek et al., 2009; Barrow & Kwon, 2009) and was shown to be involved in the biogenesis of the outer membrane (Macfarlane et al., 1999; Masschalck et al., 2003). The antibiotic polymyxin B binds to the cell membrane and thus alters its structure and permeability. These characteristics suggest that PhoPQ is responsible for sensing the temperature down-shift and together with MucA/AlgN regulating the mechanisms of cell wall maintenance.

The major cold shock protein CspA and its paralogues belonging to the *csp* family in *E. coli*, are small nucleic acid-binding proteins. Out of the nine paralogues, only four CspA, CspB, CspG and CspI are induced in response to temperature-downshifts, of which CspA was found to be essential in the acclimatization phase after cold shock (Nakashima et al., 1996; Wang et al., 1999; Phadtare & Inouye, 2004). The nucleic acid-binding domain, also called cold shock domain (CSD) is the most conserved nucleic-acid binding domain among prokaryotes and eukaryotes (Yamanaka et al., 1998) emphasizing its global importance for bacterial fitness. The cold induction of *cspA* was found to be posttranscriptional controlled. The mRNA undergoes a temperature-dependent structural conformation resulting in a stabilized transcript at low temperatures. Thus the *cspA* mRNA is less susceptible for degradation and can be more efficiently translated (Giuliodori et al., 2010). This molecular mechanism, that mRNA is able to sense a temperature down-shift and by this to act as a thermo-sensor, was also found for other cold shock proteins (Gualerzi et al., 2003).

In this transcriptome analysis only one gene coding for a cold shock protein was identified to be induced after cold shock with all three platforms, namely *capB* (PP1099). Cap proteins belong to the class of cold acclimation proteins and were already described in psychrotrophic Pseudomonads to be induced in the acclimation process after temperature down-shifts (Michel et al., 1997; Panicker et al., 2010). Beside *capB* two other cold shock-associated genes were found to be up-regulated based on Illumina and Affymetrix transcriptome data. PP0636 and *cspD* (PP4010) were slightly induced by about 4 and 3-fold respectively, whereas another cold shock domain protein was found to be repressed, PP0985 was 4-fold down-regulated. Chaperone regulation is mostly ATP dependent. Here, a chaperone-associated ATPase (PP3316) was highly induced (50-fold) and was annotated to be organized in an operon together with PP3315-PP3318, where the other three genes are hypotheticals. However, PP3316 was also detected with all three platforms and based on Affymetrix and Illumina data the whole operon was induced upon cold shock implicating an important role in response to cold shock.

Besides cold shock proteins, RNA helicases have also been proposed to function as RNA chaperones. Here, the ATP-dependent RNA helicase (PP1868) of the DEAD box gene family was induced upon cold shock. Helicases belonging to the DEAD box family are involved in several processes including transcription and initiation of translation, RNA degradation and ribosome biogenesis. The DEAD box RNA helicase CsdA of *E. coli* was already described to enhance translation initiation (Lu *et al.*, 1999) and as well as other RNA helicases such as DbpA and RhlE to support the assembly of the 50S ribosomal subunits (Elles *et al.*, 2009; Jain, 2008; Charollais *et al.*, 2004). A study by Hunger *et al.* (2006) furthermore supported a direct interaction of the cold shock protein CspB and the DEAD box helicase CshB as a double mutant affected in both genes was not viable. As FRET analysis indicated an active process on the mRNA level, they proposed a model where CshB was responsible for destabilization of mRNA secondary structures and thus providing unwound mRNA that can then be bound by CspB to prevent refolding and ensure translation initiation. PP1099, the homologous gene to *cspB* in *P. aeruginosa*, was also found to be induced upon cold shock. Though PP1099 and PP1868 are not well characterized in *P. putida*, it is likely that here a similar interaction of the RNA helicase and CspB occurs in response to cold shock.

Cellular folate is the central compound of C_1- metabolism that is the basis of DNA synthesis and *S*-adenosylmethionine-dependent methylation of proteins and DNA. In response to cold shock, *P. putida* KT2440 seemed to accumulate the folate cofactor 5,10-methenyltetrahydrofolate (5,10-methenyl-THF) by inducing the production of 5,10-methenyl-THF from 5-formyl-THF (PP5203) but repressing its conversion into either 5,10-methylene-THF and 10-formyl-THF (*folD-2*) or 5,6,7,8-THF (*purN*). This provides a storage pool for the folate metabolism and subsequent purine *de novo* synthesis once the cells have adapted to the low temperature. This was underlined furthermore by the induction of PP2512 (encoding *folE*) that catalyzes the first step in the synthesis of folate from GTP.

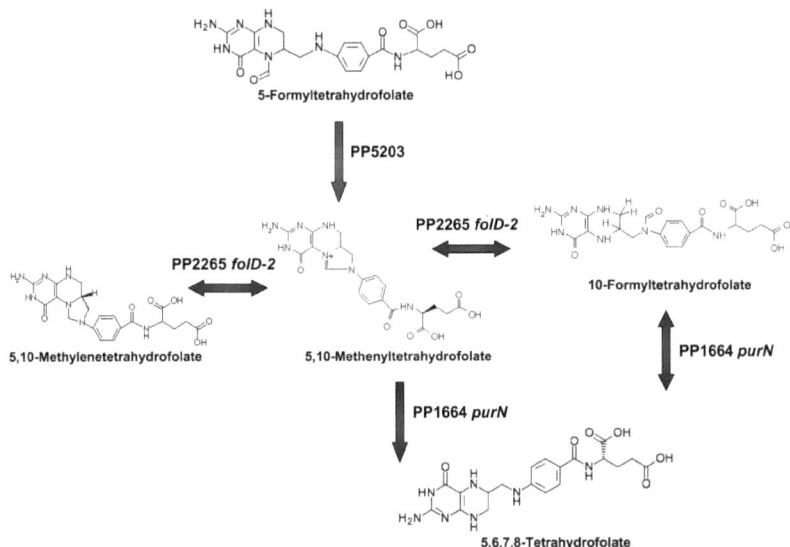

Fig. 3.25 Schematic overview of significant regulated genes involved in the pathway of one carbon pool by folate. The gene PP5203 converting 5-Formyltetrahydrofolate (5-Formyl-THF) into 5,10-Methenyl-THF was up-regulated whereas genes converting 5,10-Methenyl-THF into either 5,10-Methylene-THF, 10-Formyl-THF or 5,6,7,8-THF were up-regulated leading to an accumulation of 5,10-Methenyl-THF.

Ferritins are iron storage proteins that are either haem-free or contain haem. The latter are called bacterioferritins. Iron is essential for cell growth and its ions are involved in important processes such as respiration, DNA synthesis and nitrogen fixation and can be found in all three kingdoms of life. Bacterioferritin was already shown to be essential for iron storage and iron-mediated oxygen stress in several species such as *E. coli* (Touati *et al.*, 1995), *Campylobacter jejuni* (Wai *et al.*, 1996) and *Neisseria gonorrhoeae* (Chen & Morse, 1999). However, iron homeostasis seems to be directly affected by cold shock. A study by Phadtare & Inouye (2004) showed that in wild type cells of *E. coli* iron transport was repressed whereas in a quadruple-deletion strain affected in genes encoding four cold shock proteins (CspA, CspB, CspG and CspE) iron transport was induced. Though no genes coding for iron transport systems were detected in this study, the gene *bfr* encoding bacterioferritin was induced as in the study by Phadtare & Inouye. Since cell growth slows in response to cold shock, and metabolic processes require iron, iron storage might become essential for preventing iron-mediated oxidative stress and providing an iron "pool". This might work in a similar way to folate storage, for example during DNA synthesis, once the cells have been adapted to the decreased temperature.

RESULTS AND DISCUSSION

As the transcriptional profile of *P. putida* in response to cold shock revealed many genes encoding hypothetical proteins, several of which were analyzed in more detail with bioinformatic tools to search for conserved domains and motifs that might indicate a functional role of the respective genes. The following programs were included in the analysis of encoded protein sequence:

- Blastp using the PSI-BLAST (Position-Specific Iterated BLAST) algorithm (http://blast.ncbi.nlm.nih.gov/Blast; Altschul *et al.*, 1990) to find distantly related proteins stored in the database.
- Pfam to find conserved domains that might give insights into the function of the protein (http://pfam.sanger.ac.uk/; Finn *et al.*, 2008).
- HAMAP is generally used for annotation of proteins by searching for well-defined protein families or subfamilies (http://www.expasy.ch/sprot/hamap/index.html; Gattiker *et al.*, 2003) and is linked to the UniProtKB/Swiss-Prot protein database (http://www.uniprot.org/; Wu *et al.*, 2006).
- TmHMM for the prediction of transmembrane regions (http://www.cbs.dtu.dk/services/TMHMM/; Krogh *et al.*, 2001).
- PSORTp for the prediction of subcellular location (http://www.psort.org/psortb/; Gardy *et al.*, 2005).
- KEGG for the identification of possible pathways in which the encoded enzyme might be involved (http://www.genome.jp/kegg/; Kanehisa *et al.*, 2008).

In total, eight genes or rather the encoded protein sequences were included in the analysis. PP5232 was strongly down-regulated according to the Illumina results (138-fold), PP1147 and PP1148 which are organized in an operon were eight and 15-fold up-regulated, respectively, and PP1690 and PP1691 which are together with PP1689 organized in an operon. PP1690 and PP1619 were the genes that displayed the highest up-regulation upon cold shock, 487 and 1232-fold with the Illumina sequencing and 279 and 166-fold with the Affymetrix microarrays, respectively. Furthermore, PP2296 and PP3317 were also included in the analysis.

No conserved domain could be found for five out of the eight genes, PP1147, PP1148, PP1690, PP1691 and PP5232, using any of the prediction programs, of which only for one gene a prediction for cellular localization could be made. PP1690 was predicted to be associated with the cytoplasmic membrane with a significant localization score of 9.86 using PSORTp. Furthermore, the remaining four genes seem to be conserved within the genus *Pseudomonas* as no homologues were found within other genera.

RESULTS AND DISCUSSION

The gene PP2296 has no orthologues in other *Pseudomonas* species, but displayed some similarity to transcriptional regulators annotated in *Azospirillium* and *Rhodospirillium* species with e-values of 9e-11 and 6e-07, respectively. The Pfam prediction program revealed a domain with minor similarities (e-value 0.1) to Salt_to_Pase (PF09506). This domain is found in protein families belonging to with the gene name *stpA* (salt tolerance protein A). The glucosylglycerol phosphate phosphatases StpA was first identified in the *Synechocystis*, a moderately halo-tolerant cyanobacterium (Hagemann *et al.*, 1996, 1997), where it is involved in the conversion of glucosylglycerol phosphate, the precursor of the osmoprotectant glucosylglycerol (Borowitzka *et al.*, 1980). A *stpA*-negative mutant was salt sensitive and accumulated glucosylglycerol phosphate. Osmotic stress caused by high salt concentrations leads to loss of water and increasing intracellular ion concentration. One adaptive response mechanism to osmotic stress is the altered regulation of transporter systems to either extrude excessive ions or to enhance the uptake of compatible solutes. The transcriptome data already indicated that a transporter system was differentially regulated, that is likely to be involved in ectoine uptake, a compatible solute that was found to be important in response to osmotic stress (García-Estepa *et al.*, 2006; Bursy *et al.*, 2008; Naughton *et al.*, 2009) and heat stress (Harishchandra *et al.*, 2010) in various species. These studies showed already that the mechanisms in response to osmolarity or unfavourable temperatures are related as the uptake of compatible solutes is important for cell adaptation. Hence, PP2296 might be involved in the synthesis of compatible solutes in *P. putida* in response to decreasing temperatures as indicated by the identified domain, though there is no further evidence for this from the other prediction programs.

PP3317 is predicted to be organized in an operon together with PP3315, PP3316 and PP3318 whereof only PP3316 has a predicted function as a chaperone-associated ATPase, as already mentioned earlier in this chapter. Since they are associated with PP3317, the remaining three genes of this operon were also included in the analysis. PP3316 belongs to the AAA family proteins (PF00004) that are involved in the assembly and disassembly of protein complexes, thus displaying chaperone-like functions (Neuwald *et al.*, 1999). It contains furthermore a Clp amino terminal domain (PF02861) that is found in the amino terminus region of the ClpA and ClpB proteins and is likely to act as a protein binding site (Confalonieri & Duguet, 1995). In general Clp ATPases are involved in protein degradation, though ClpB in particular is associated with protein disaggregation (Lee *et al.*, 2003). According to PSORTp prediction it is localized in the cytoplasma. PP3317, which was induced 11.5-fold upon cold shock, has no orthologues in other *Pseudomonas* species. The Blast search revealed some similarity to the small heat shock protein HspC2 of *Achromobacter piechaudii* with an e-value of 3e-04. Indeed, the Pfam prediction program found a conserved domain, Hsp20 (PF00011). Proteins containing this domain seem to act as chaperones in heat shock response by ensuring protein assembly and functionality (Lindquist & Craig, 1988; Kim et al., 1998). PP3318 like PP3317 has no orthologues within the genus *Pseudomonas*. The Pfam prediction found a domain belonging to the family of ribosomal proteins S14p (PF00253). Ribosomes are the catalysts in mRNA-directed protein synthesis. The S14 ribosomal protein belongs to the small ribosomal subunit. In *E. coli* it was described to be involved in the assembly of the 30S complex and to be responsible for 16S rRNA conformation at the A site (Yuki & Brimacombe, 1975; Ramakrishnan et al., 1984). PP3315 was predicted to contain four transmembrane helices and to be localized in the cytoplasmic membrane. Furthermore, the phosphate-starvation-inducible E domain (PF06146) was found. In *E. coli* the phosphate-starvation-inducible gene *psiE* belongs to the pho regulon whose expression is induced due to phosphate starvation and whereof some genes are involved in the transport and degradation of phosphorus sources (Kim et al., 2000). Considering the identified domains of all four genes belonging to the operon, as reflected by the induction of PP3316 and PP3317, is an important gene cluster involved in protein fate by enhancing translation efficiency and assisting in ribosome assembly as indicated by PP3318, which belongs to the ribosomal protein S14 protein family of. It also might be involved in the degradation of dysfunctional proteins as indicated by the domains found in PP3316 and PP3315. In response to low temperatures, ribosome activity and thus protein synthesis is hampered. Hence, this operon PP3315-PP3318 might be a major protein complex that exhibits chaperone-like functions.

In total, 159 genes had been identified to be differentially regulated upon cold stress. One finding was the high abundance of encoded hypothetical proteins among the induced genes, 38%, in contrast to 7% among the repressed genes. While the hypothesis that the repressed genes generally represented a down-regulation of intermediary metabolism and regulatory mechanisms was well supported by existing literature, the set of genes reflecting the major mechanisms that are induced upon cold shock were hardly interpretable. The bioinformatic analysis of selected differentially expressed hypothetical proteins revealed that much remains to be elucidated in the cold shock response of *P. putida* as many genes that are the key players in cold shock response are still relatively unknown, making a functional prediction difficult. For five out of the eight genes that were analysed in more detail, no functional prediction could be made at all, since no conserved domains or motifs were found. For the remaining three genes, the functional predictions were rather limited, since the confidence values were very low. However, it is worthwhile to mention that particularly these genes seem to be highly conserved within the genus *Pseudomonas*, demonstrating once more that little is known about the genes and the up-regulated mechanisms upon cold shock in Pseudomonads.

3.7.2 Proteome analysis

The transcriptional profiling of the cold shock response was supplemented with a proteomic approach. For this, samples derived from the same bacterial cultures as for transcriptome analysis were prepared and sent to the Department of Functional Genomics at the Ernst-Moritz-Arndt-University of Greifswald for further processing, mass spectrometric analysis and preliminary data evaluation.

The proteome data showed overall a weak differential expression. In agreement with the proteome facility, detected proteins were considered to be significantly regulated if for both biological samples the calculated p-value was ≤ 0.05 and regarded as differentially regulated if fold-change ≥ 1.5.

In total, 203 proteins were identified that fulfilled the statistical cut-off, of which 59 were down or up-regulated by more than 1.5-fold. The 59 proteins are listed in Table 3.17. represented by the statistical mean value of both biological samples.

The overrepresented portion (83%) of up-regulated proteins correlated very well with the percentaged number of genes identified by Illumina and Affymetrix where more than two-thirds of total genes were induced upon cold shock (see Table 3.15).

Table 3.17 Summary of proteins that were detected in both biological samples with a calculated p-value ≤ 0.05 (T-test) and an absolute fold-change of at least 1.5.

	locus id	fold-change[§]	p-value[§]	protein	protein name
*	PP0268	[#]1.58	0.0120	outer membrane protein OprE3	OprQ
A	PP0566	-1.66	0.0460	translation initiation factor SUI1	
	PP0572	3.31	0.0010	penicillin-binding protein 1C	PbpC
	PP0623	1.53	0.0240		
AI	PP0773	1.54	0.0003	OmpA family protein	
	PP0799	1.82	0.0006	porin, putative	
AI	PP0857	1.54	<0.0001	GTP-binding protein EngA	EngA
*	PP0868	2.68	<0.0001	glycine betaine/carnitine/choline ABC transporter, AT	
*	PP0870	5.19	<0.0001	glycine betaine/carnitine/choline ABC transporter, pe	
AI	PP0886	1.96	<0.0001	conserved hypothetical protein	
A	PP1024	1.54	0.0320	2-dehydro-3-deoxyphosphogluconate aldolase -	Eda
*	PP1071	-2.86	<0.0001	amino acid ABC transporter, periplasmic amino acid-binding	
*	PP1082	[#]-1.81	0.0020	bacterioferritin	Bfr
*	PP1099	1.70	<0.0001	cold shock domain family protein	
*	PP1111	-1.80	0.0090	synthetase, putative	
	PP1131	1.76	0.0001	outer membrane lipoprotein, putative	
AI	PP1141	-1.90	<0.0001	branched-chain amino acid ABC transporter, periplasmic	BraC
	PP1185	1.68	<0.0001	outer membrane protein H1	OprH
*	PP1186	1.52	0.0001	transcriptional regulatory protein PhoP	PhoP
	PP1223	1.57	0.0010	peptidoglycan-associated lipoprotein OprL	OprL
AI	PP1251	1.60	0.0008	malate:quinone oxidoreductase	Mqo-2
*	PP1428	[#]-2.38	<0.0001	sigma factor algU negative regulatory protein MucA	MucA
AI	PP1433	1.76	<0.0001	ribonuclease III	Rnc
	PP1463	1.73	<0.0001	16S rRNA processing protein RimM	RimM
I	PP1835	1.75	0.0050	conserved hypothetical protein	
*	PP1868	3.67	<0.0001	ATP-dependent RNA helicase, DEAD box family	
I	PP2089	[#]1.54	<0.0001	outer membrane protein OprF	OprF
	PP2105	1.73	0.0320	conserved hypothetical protein	
*	PP2296	1.66	<0.0001	hypothetical protein	
	PP2322	2.42	<0.0001	outer membrane lipoprotein OprI	OprI
I	PP2396	-1.67	0.0220	hypothetical protein	
AI	PP2448	9.07	<0.0001	conserved hypothetical protein	
	PP2466	1.54	<0.0001	translation initiation factor IF-3	InfC
AI	PP2936	[#]1.85	<0.0001	ABC transporter, ATP-binding protein	
AI	PP3930	2.35	<0.0001	hypothetical protein	
AI	PP4400	1.71	0.0060	cell division protein FtsK	FtsK
	PP4378	-3.30	<0.0001	flagellin FliC	FliC
AI	PP4470	2.23	<0.0001	alginate biosynthesis transcriptional activator	AlgZ
	PP4496	2.03	<0.0001	hypothetical protein	
*	PP4548	1.53	0.0020	oxidoreductase, putative	
I	PP4563	1.72	0.0010	conserved hypothetical protein	
	PP4591	1.51	0.0001	ribonuclease D	Rnd
I	PP4683	1.85	0.0090	penicillin-binding protein	MrcB
AI	PP4711	1.55	<0.0001	ribosome-binding factor A	RbfA
	PP4787	2.38	<0.0001	PhoH family protein	
A	PP4788	1.90	0.0009	conserved hypothetical protein TIGR00043	
I	PP4809	1.68	<0.0001	conserved hypothetical protein	
*	PP4870	-1.56	0.0010	azurin	
	PP4879	1.69	<0.0001	RNA methyltransferase, TrmH family, group 3	
	PP4880	1.64	<0.0001	ribonuclease R	VacB
	PP4974	1.51	0.0500	Na+/H+ antiporter, putative	
A	PP5038	1.52	0.0370	conserved hypothetical protein	
	PP5044	1.51	<0.0001	GTP-binding protein TypA/BipA	
I	PP5114	1.61	0.0090	conserved hypothetical protein TIGR00095	
AI	PP5172	-2.69	0.0001	conserved hypothetical protein	
A	PP5184	1.52	<0.0001	glutamine synthetase, putative	
	PP5187	1.55	<0.0001	conserved hypothetical protein	
AI	PP5278	1.86	<0.0001	aldehyde dehydrogenase family protein	
I	PP5338	1.60	<0.0001	aspartate ammonia-lyase	AspA

[§] Fold-change and p-value were calculated as geometric mean value derived from both biological samples.
[#] Discrepancy in regulation compared to transcriptome data.
[A] Identified additionally by Affymetrix microarrays.
[I] Identified by Illumina cDNA sequencing.
[AI] Identified by Affymetrix microarrays and Illumina cDNA sequencing
[*] Detected with proteome analysis and all three transcriptome platforms.

Out of the 59 proteins, 18 were identified only by the proteomics approach; the corresponding genes were not detected in the transcriptome profiling. This result, that about one-third of proteins showed no significant differential expression on the transcription level supports the approach of integrating different "omics" for a complementary view of the biological processes involved. Another 28 proteins were identified by at least two transcriptome platforms (15 identified by Affymetrix and Illumina, 13 identified by all three platforms), the remaining 13 were detected by either Affymetrix or Illumina.

One major impact of cold shock is the formation of mRNA secondary structures and the resultant constraint on translation. This problem can be circumvented by synthesizing cold shock proteins (Michel et al., 1997; Weber & Marahiel, 2002; Panicker et al., 2010) and RNA helicases that act as RNA chaperones and help to unwind stable mRNA structures (Hunger et al., 2006) respectively to enhance translation efficiency (Lu et al., 1999) and assembly of 50S ribosomal subunits (Elles et al., 2009; Jain, 2008; Charollais et al., 2004). Appropriate protein amounts and synthesis in the adaptation phase after cold shock are dependent on translational efficiency, fidelity and on available mRNA transcripts.

In the proteomic profiling 12 proteins could be assigned to processes involved in transcription and translation that were all induced upon cold shock. Six out of these 12 proteins are required for ribosome assembly. RimM and RbfA, the ribosome binding factor, are essential for 16S rRNA processing. Furthermore, two GTP-binding proteins EngA and BipA and two translation initiation factors, SUI1 and IF-3, were identified by the proteome data, indicating the crucial role of ribosome maintenance in recovering translation efficiency (Lockwood et al., 1972; Bollen et al., 1975; Cairrão et al., 2003; Hong et al., 2005). Proteins involved in RNA processing were found to be induced, such as ribonucleases III and D, which play a general role in transcription. Others included an RNA methyltransferase (PP4879) and VacB, a ribonuclease R, as well as PP1868, the ATP-dependent RNA helicase that is involved in RNA modification and degradation (O'Hara et al., 1995; Cheng & Deutscher, 2003, 2005).

The identification of the ATP-dependent RNA helicase together with the cold shock protein CapB (PP1099) in both transcriptome and proteome data supports the hypothesis that these two proteins play a major role in cold shock response in *P. putida*.

GTP-binding proteins are furthermore involved in signal transduction as GTP is a key signalling molecule being the precursor of the regulatory compound ppGpp (Cashel & Gallent, 1969; Haseltine & Block, 1973), and are generally known to mediate peptide chain initiation and elongation during translation (Moore, 1995). However, they were also described to be involved in alginate synthesis in *P. aeruginosa* as GTP is a precursor of GDP-mannose (May et al., 1994; Sundin et al., 1996). As the GTP-binding proteins were found to be induced on the protein level, it seems likely that these proteins are involved in sensing the temperature down-shift and in activating intracellular mechanisms to cope with cold-stress induced phenomena, such as a decrease in membrane fluidity by activating alginate synthesis and countering hampered translation efficiency by enhancing its initiation.

The role of alginate production for the maintenance of membrane fluidity was already discussed in the previous section, but was additionally confirmed by the induction of the protein AlgZ, an alginate biosynthesis transcriptional activator, controlling the expression of *algD* and by this alginate synthesis. The proteome and transcriptome data showed both the induction of one alginate transcriptional regulator, AlgZ and AlgN respectively. These data emphasize the important role of cell wall maintenance by alginate biosynthesis. They further indicate that the alginate transcriptional regulators together with PhoP and MucA, that were found to be induced with both approaches, build the main regulatory entity in cold shock response.

Furthermore, a high number of membrane-associated proteins (16) such as transporter systems, porins, outer membrane proteins (OprQ/H/L/F/I) and two penicillin-binding proteins, PbpC and MrcB, which are involved in peptidoglycan biosynthesis, were detected to be differentially expressed in response to cold shock. The last subunit of the ABC transporter (PP1068-1071) was found to be repressed upon cold shock. This transporter was already identified by the transcriptome data supporting the idea that ectoine (as the first subunit was annotated to be involved in ectoine transport) acts as a compatible solute in cold shock response and is thereby involved in repair mechanisms of the cell wall by increasing the fluidity of the membrane (Harishchandra et al., 2010). The importance of the uptake of compatible solutes was furthermore emphasized by the induction of the glycine betaine ABC transporter (PP0868-0870) that was also identified by the transcriptome analysis. PbpC and MrcB mediate the last step in peptidoglycan biosynthesis which is the major compound of the bacterial cell wall.

In five out of the 59 proteins a discrepancy in regulation between the transcriptome and proteome data was observed, in two cases the respective genes were induced upon cold shock but the proteins were down-regulated according to the proteome data whereas 3 genes showed a repressed expression with the corresponding proteins found to be up-regulated. Among these five proteins, four were broadly associated with cell wall biogenesis. These four include MucA, OprQ, OprI and PP2936, encoding an ABC transporter subunit, where the latter three were induced on the protein level. This finding, together with the considerable number of proteins exhibiting differential regulation on the protein level whose corresponding genes were not found to be expressed due to the transcriptome profiling, indicate that post-transcriptional regulatory mechanisms are essential in the adaptation process after cold shock and strongly influence the protein pattern of experiencing cold shock.

The differences of the transcriptome and proteome profile were furthermore emphasized by comparing the gene and protein lists according to the respective pathways or protein groups. The genes or proteins identified to be significantly differentially expressed with all three transcriptome platforms or by the proteome approach as well as the genes identified by Illumina cDNA sequencing were assigned to functional categories based on the KEGG annotation of *P. putida* KT2440. The genes were either assigned to regulatory pathways or processes (see Figure 3.26) or functional classes of proteins (see Figure 3.27).

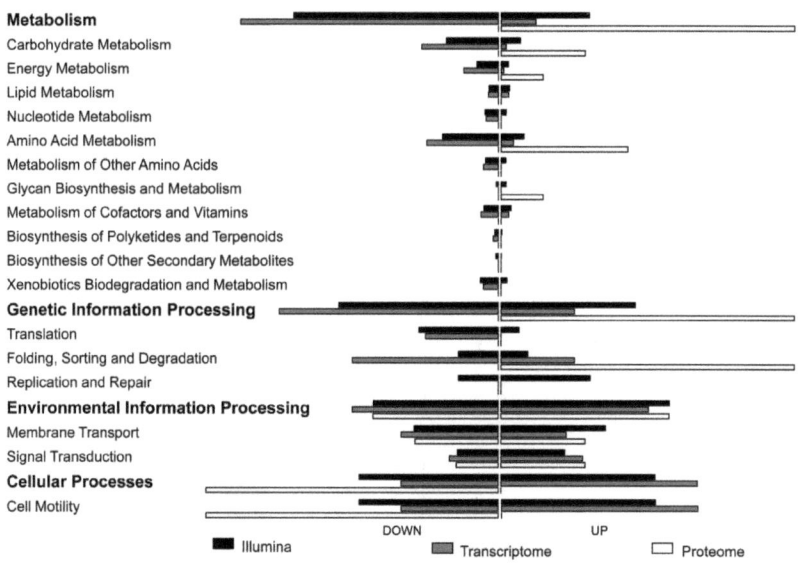

Fig. 3.26 Classification of cold shock response of *P. putida* KT2440 to functional categories. Each plot indicates the type of physiological roles and the percentage of assigned genes and proteins whose expression increased or decreased upon cold shock that were identified by either Illumina cDNA sequencing alone, transcriptome analysis including all three platforms or the proteome approach.

Out of the 2337 and 159 genes that were found by Illumina cDNA sequencing or consistently with all three platforms, 1730 and 93 could not be assigned to any regulatory pathway or process, and 1507 and 76 not to any functional class of proteins, respectively. Of the 59 proteins that were significantly differentially expressed, 43 and 35 could not be matched to either a regulatory pathway or a functional protein class. This demonstrated once more that many genes and proteins belonging to the cold shock regulon of *P. putida* KT2440 are still unknown as regards their regulatory function.

For the remaining genes and proteins, the number of assigned genes (proteins) found for each pathway or functional class of proteins was expressed in percentage according to the total number of assigned genes (proteins) per list. Though only a few proteins could be assigned, the comparison of the transcriptome and proteome approach showed significant differences. In the functional categories belonging to either metabolism or genetic information processing, only proteins were found that were induced upon cold shock, none of the repressed proteins could be assigned to either of the categories. In contrast, pathways reflected by genes that were identified by transcriptome analysis were mainly down-regulated as illustrated by carbohydrate, energy and amino acid metabolism. Within the categories that belong to the intermediary metabolism, lipid and nucleotide metabolism, metabolism of cofactors and vitamins and xenobiotics degradation and metabolism were only represented by genes identified by the transcriptome analysis, but no protein was found in either of these groups. In contrast, glycan biosynthesis and metabolism was mainly reflected by identified proteins, and only a few genes that were identified by Illumina sequencing could be assigned to this group. Within categories belonging to genetic information processing, proteins could only be assigned to the category of folding, sorting and degradation whereas genes derived by the transcriptome were distributed over all three categories, translation, folding, sorting and degradation and replication and repair. Hereby, the functional category of translation was mainly reflected by genes that were down-regulated upon cold shock. The finding that pathways represented by the identified proteins were mainly up-regulated correlated well with the total number of induced proteins as 83% of significantly differentially regulated proteins were indeed induced. In contrast, more genes that were repressed upon cold shock were assigned to pathways of the intermediary metabolism, though overall more genes were up-regulated upon cold shock according to the Illumina sequencing results (63.2%). The transcriptome data as here displayed by the functional categories therefore reflected broadly the general down-regulation of intermediary metabolism as previously reported in response to decreasing temperatures.

RESULTS AND DISCUSSION

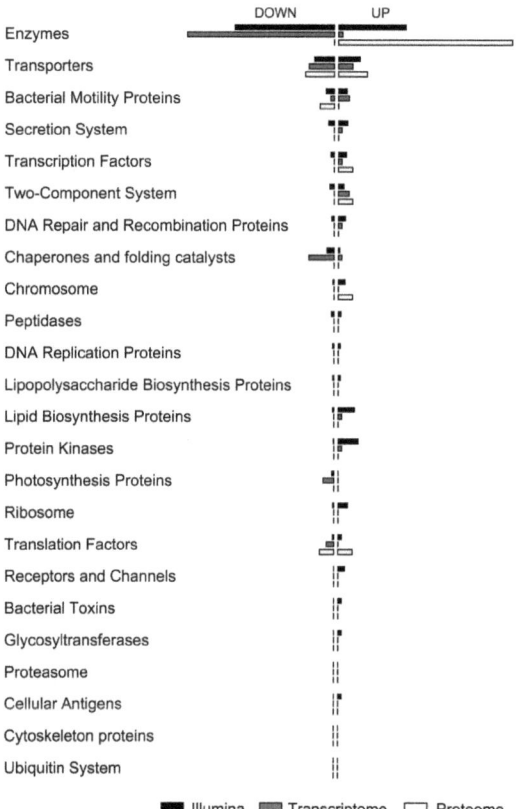

Fig. 3.27 Classification of cold shock response of *P. putida* KT2440 to functional classes of proteins. Each plot indicates the functional role and the percentage of assigned genes and proteins whose expression increased or decreased upon cold shock that were identified by either Illumina cDNA sequencing alone, transcriptome analysis including all three platforms or the proteome approach.

In addition, the same gene and protein lists were assigned to functional classes of proteins. The functional class of enzymes, that exhibited the highest number of assigned genes and proteins, showed major differences. The genes identified by Illumina were similarly distributed between the up and down-regulated enzymes whereas only up-regulated proteins could be assigned to this group and nearly all genes identified by all three platforms that act as enzymes were down-regulated. Besides enzymes, transporters, bacterial motility proteins, transcription factors, two-component systems, chaperones and folding catalysts and translation factors were the major functional classes of proteins important for cold shock response. Transporters were the second largest group with an equal distribution of up and down-regulated genes (proteins), thus emphasizing that the regulation of transporter systems is one of the major mechanisms to alter the cell membrane composition in response to decreasing temperatures. The data furthermore showed that transcription factors and two-component systems are mainly up-regulated in response to cold shock, identified by transcriptome and proteome. Furthermore, two functional classes of proteins were mainly reflected by indentified proteins, namely chromosomes and translation factors. As samples were taken two hours after the temperature of the medium reached 10°C, the induction of genes encoding translation factors or proteins related to chromosome might be the initial regulatory mechanism starting directly in response to the decreasing temperature. Thus, the regulation of these mechanisms was no longer detectable on the gene expression level, but on the protein level. The same was observed for the glycan biosynthesis and metabolism pathway that were mainly reflected by induced proteins whereas no genes from the comparative transcriptome analysis and only few genes derived from the Illumina results could be assigned to this functional category. This is agreement with the general regulatory mechanisms to cold shock response as the formation of stable mRNA secondary structures leads to reduced translational efficiency and the membrane fluidity is impaired at low temperatures (Phadtare, 2004). Thus, the regulation of translation factors to enhance translation efficiency and the induction of the glycan biosynthesis pathway for the production of cell membrane compounds represent the first altered mechanisms in cold shock response.

According to the assigned genes, the functional classes of chaperones and folding catalysts are mainly down-regulated. This finding was interesting since cold and heat shock proteins act as chaperones that are induced or repressed upon temperature shift. The transcriptome data showed that *groEL* and *groES* are highly induced upon cold shock, which have already been described to act either as heat shock proteins (Bardwell & Craig, 1984; Holtmann *et al.*, 2004; Carruthers & Minion, 2009) or as cold shock proteins (Phadtare & Inouye, 2004). In contrast *dnaK* and *grpE* are well characterized heat shock proteins (Dubern *et al.*, 2005; Firoved *et al.*, 2002) which were repressed. However, the strong tendency of down-regulated genes assigned to the chaperone functional class suggests that the major cold shock proteins in *P. putida* are not identified yet, as these should be induced in response to cold shock.

In summary, the combination of a transcriptome and proteome approach revealed the key players of cold shock stress in *P. putida* KT2440. The proteome approach complemented the transcriptome data by detecting additional proteins involved in the same biological processes, mainly cell wall biogenesis to maintain membrane fluidity and translation by circumventing translational constraints due to hampered ribosome function and secondary structures of mRNA. Beside this, the considerable number of proteins (18 of 59) that displayed differential regulation, but were not identified by the transcriptional profiling, emphasizes that post-transcriptional regulation contributes strongly to the survival and adaptation of *P. putida* KT2440 after cold shock. Furthermore, though the here described mechanisms reflected by the identified differentially regulated genes and proteins are generally known as described in the literature, the data here emphasizes once more that cold shock response is distinct in different species as exemplified by the differential regulation of genes in response to a temperature down-shift, e.g. *groEL* and *groES* in *E. coli* and *P. putida*. The highest induction on the transcriptional level upon cold shock was observed for PP1691 (1230-fold) which is organized in an operon together with PP1689 and PP1690 according to the Illumina cDNA sequencing results. These three genes have no predicted function yet, but indicated by the high expression of all three genes, it is likely that this gene cluster plays a key regulatory role in cold shock response. The assignment of identified genes and proteins revealed indeed that the major cold shock proteins have not yet been identified in *P. putida* as the functional class of chaperones was constituted mainly by down-regulated genes, in contrast to expectation.

It is worthwhile to mention that two of the five cold sensitive mutants were found with the proteomics approach, but none of the mutants were detected with the transcriptome analysis; VacB and BipA were 1.6 and 1.5-fold up-regulated according to the proteome data, demonstrating their important role in cold shock response.

3.8 The Transcriptome Profile of Cold Sensitive Tn5 Mutants

The cold sensitive mutants used in this study, previously identified by Reva *et al.* (2006) were verified to check correct geno- and phenotype (see chapter 3.1 and 3.2).

The two-component system CbrAB and the poly(A) polymerase PcnB are located within the same genomic region and were already described to constitute a regulatory unit, as *pcnB* expression is dependent on CbrB (Zhang *et al.*, 2010). The exoribonuclease R VacB is involved in RNA processing (Li *et al.*, 2002; Cheng & Deutscher, 2005; Andrade *et al.*, 2006) and was described to be important for growth at low temperatures (Cairrão *et al.*, 2003; Erova *et al.*, 2008; Tsao *et al.*, 2009). The GTPase protein BipA exhibits several ribosome-associated cellular functions related to stress response (Farris *et al.*, 1998; Pfennig & Flower, 2001; DeLivron *et al.*, 2009) (see also chapter 1.2).

Similar to the cold shock experiments with *P. putida* KT2440 wild type, samples from the same cultures for each mutant were used for subsequent RNA preparation or intracellular metabolite extraction for a comprehensive analysis of the transcriptome and metabolome profile of the respective mutants after cold shock in comparison to the wild type. As the transcriptome profile of cold sensitive mutants was to be compared to the wild type profile, cDNA samples from the 10°C culture after cold shock of wild type and respective mutants were hybridized onto one microarray. The difference in gene expression thus reflected the distinct cold shock response of the respective mutants. For metabolome analysis, samples from both 30°C and 10°C for each strain were processed.

For the transcriptome analysis of the five cold sensitive mutants (deficient in *cbrA*, *cbrB*, *pcnB*, *vacB* or *bipA*), genes that were found to be significantly differentially expressed in both biological samples were considered to be representative for the transcriptional profile of the respective mutants. A list of differentially expressed genes for each mutant is provided in Table 3.18. The statistical parameters, p-value (two-sample T-test) and FDR as well as absolute fold-change were calculated as geometric mean derived from the two biological samples.

Table 3.18   Summary of differentially expressed genes* in cold sensitive transposon mutants in comparison to the wild type transcriptome profile at 10°C.

locus id	cbrA::Tn5			cbrB::Tn5			pcnB::Tn5			vacB::Tn5			bipA::Tn5			encoded protein	gene name
	p-value	FDR	fold-change	p-value	geom. FDR	fold-change	p-value	FDR	fold-change	p-value	FDR	fold-change	p-value	FDR	fold-change		
PP0018				0.0008	0.0369	-2.66										hypothetical protein	
PP0021				0.0014	0.0510	-2.92										hypothetical protein	
PP0038				0.0003	0.0233	-3.25										hypothetical protein	
PP0117										0.0001	0.0066	2.04				zinc ABC transporter, permease protein ZnuB	znuB
PP0153				<0.0001	0.0034	-10.31										conserved hypothetical protein	
PP0255										0.0008	0.0266	-2.58				conserved hypothetical protein	
PP0275													0.0005	0.0431	-2.45	conserved hypothetical protein	
PP0276										0.0003	0.0166	-2.26				transcriptional regulator, Cro/CI family	
PP0317										<0.0001	0.0047	-2.34				methyl-accepting chemotaxis transducer	
PP0320										0.0002	0.0148	-1.92				methyl-accepting chemotaxis transducer	
PP0362				0.0001	0.0090	2.74										biotin synthetase	bioB
PP0536				0.0004	0.0259	-4.12				0.0001	0.0069	-2.49	<0.0001	0.0065	-3.63	hypothetical protein	
PP0584				0.0001	0.0099	-5.24				0.0012	0.0385	1.84				methyl-accepting chemotaxis transducer	
PP0596													0.0018	0.0761	-2.03	beta-alanine--pyruvate transaminase	
PP0600										<0.0001	0.0024	-2.52				ribosomal protein S20	rpsT
PP0721				0.0003	0.0208	-3.43										ribosomal 5S rRNA E-loop binding protein Ctc/L25/TL5	
PP0816				0.0004	0.0244	-2.64										protoheme IX farnesyltransferase	cyoE-2
PP1009				0.0012	0.0487	-2.80										glyceraldehyde 3-phosphate dehydrogenase	gap-1
PP1011				0.0006	0.0325	-3.06										glucokinase	glk
PP1075				0.0007	0.0363	-2.67										glycerol kinase	glpK
PP1116													0.0007	0.0501	-2.32	site-specific recombinase, resolvase family	
PP1148				0.0038	0.0935	-2.53										hypothetical protein	
PP1185				0.0003	0.0212	-3.40										outer membrane protein H1	oprH
PP1186				<0.0001	0.0076	-5.34										transcriptional regulatory protein PhoP	phoP
PP1187				<0.0001	0.0030	-5.62										sensor protein PhoQ	phoQ
PP1244				0.0010	0.0429	2.64										hypothetical protein	
PP1248													0.0015	0.0690	1.93	transporter, LysE family	
PP1305				0.0002	0.0152	3.07							<0.0001	0.0022	3.20	Pyocin S-type immunity protein, putative	
PP1328				0.0029	0.0775	-3.14										conserved hypothetical protein TIGR00242	
PP1330				0.0006	0.0331	-2.38										cell division protein FtsL	ftsL

Table 3.18 continued

locus id	cbrA::Tn5 p-value	cbrA::Tn5 FDR	cbrA::Tn5 fold-change	cbrB::Tn5 p-value	cbrB::Tn5 FDR	cbrB::Tn5 fold-change	pcnB::Tn5 p-value	pcnB::Tn5 FDR	pcnB::Tn5 fold-change	vacB::Tn5 p-value	vacB::Tn5 FDR	vacB::Tn5 fold-change	bipA::Tn5 p-value	bipA::Tn5 FDR	bipA::Tn5 fold-change	encoded protein	gene name
PP1334										0.0002	0.0103	2.01				phospho-N-acetylmuramoyl-pentapeptide-transferase	mraY
PP1668				0.0014	0.0538	-3.21										DnaA family protein	
PP1670				0.0032	0.0840	-2.45										lipoprotein, putative	
PP1689				0.0003	0.0200	-5.58										long-chain fatty acid transporter, putative	
PP1690				0.0001	0.0079	-5.72										conserved hypothetical protein	
PP1692				<0.0001	0.0071	-5.86										hypothetical protein	
PP1788													0.0014	0.0702	-2.13	hypothetical protein	
PP1789													<0.0001	0.0068	-3.01	hydrolase, haloacid dehalogenase-like family	
PP1798				0.0005	0.0316	-2.59										outer membrane efflux protein	
PP1859				<0.0001	0.0066	-4.95										organic hydroperoxide resistance protein	ohr
PP1977										0.0001	0.0098	-2.19				glutamyl-tRNA synthetase	gltX
PP2198										<0.0001	0.0009	-2.38				conserved hypothetical protein	
PP2256										0.0001	0.0067	-2.74	0.0003	0.0295	-2.62	transcriptional regulator, Cro/CI family	
PP2259										<0.0001	0.0050	-2.75	<0.0001	0.0081	-2.59	sigma-54 dependent transcriptional regulator	
PP2292				<0.0001	0.0017	8.16										hypothetical protein	
PP2297				<0.0001	0.0013	-7.28										integrative genetic element Ppu40, integrase	
PP2303				0.0006	0.0346	2.66										DNA-binding protein HU-beta	hupB
PP2447				0.0013	0.0491	-2.27										conserved hypothetical protein	
PP2581				0.0015	0.0554	-2.66										hypothetical protein	
PP2595										<0.0001	0.0046	-2.26				ABC transporter, permease/ATP-binding protein, putati	
PP2627										0.0014	0.0381	-2.07				conserved hypothetical protein	
PP2629													0.0006	0.0394	2.02	hypothetical protein	
PP2663				<0.0001	0.0006	-17.06										conserved hypothetical protein	
PP2664	<0.0001	0.0092	-5.07	0.0001	0.0149	-4.99	<0.0001	0.0173	-4.62	0.0003	0.0166	-2.78	<0.0001	0.0003	-4.62	sensory box histidine kinase/response regulator	
PP2665				0.0021	0.0643	-3.12							0.0002	0.0224	-2.96	DNA-binding response regulator AgmR	agmR
PP2666	<0.0001	0.0008	-6.23	0.0001	0.0111	-5.51	<0.0001	0.0256	-4.88	0.0014	0.0493	-2.13	<0.0001	0.0016	-4.63	hypothetical protein	

Table 3.18 continued

locus id	cbrA::Tn5			cbrB::Tn5			pcnB::Tn5			vacB::Tn5			bipA::Tn5			encoded protein	gene name
	p-value	FDR	fold-change	p-value	FDR	fold-change	p-value	locus id	p-value	FDR	fold-change	p-value	FDR	fold-change	p-value		
PP2673										<0.0001	0.0040	-2.65				pentapeptide repeat family protein	
PP2674										<0.0001	0.0038	-3.14				quinoprotein ethanol dehydrogenase	qedH
PP2675	<0.0001	0.0025	-5.55	<0.0001	0.0019	-6.03	0.0001	0.0369	-4.48	<0.0001	0.0062	-3.30	<0.0001	0.0013	-4.04	cytochrome c-type protein	
PP2676	<0.0001	0.0008	-8.08	<0.0001	0.0033	-7.82	<0.0001	0.0082	-7.26	0.0004	0.0238	-3.80	<0.0001	0.0003	-6.19	periplasmic binding protein, putative	
PP2677	<0.0001	0.0070	-6.03	0.0002	0.0158	-6.14	<0.0001	0.0183	-5.03				<0.0001	0.0031	-3.80	hypothetical protein	
PP2678																hydrolase, putative	
PP2679										<0.0001	0.0048	-2.63	0.0003	0.0274	-2.24	quinoprotein ethanol dehydrogenase, putative	
PP2680	<0.0001	0.0073	-6.24	<0.0001	0.0041	-6.45	<0.0001	0.0082	-6.12	0.0001	0.0083	-3.71				aldehyde dehydrogenase family protein	
PP2681	<0.0001	0.0028	-5.31	0.0003	0.0205	-4.26				0.0001	0.0081	-3.90				coenzyme PQQ synthesis protein D, putative	
PP2682	0.0002	0.0600	-2.62	0.0003	0.0224	-3.06										alcohol dehydrogenase, iron-containing	
PP2842				0.0025	0.0723	-2.33										urease accessory protein UreD	ureD
PP2927				0.0006	0.0326	2.46										conserved hypothetical protein	
PP2970				0.0002	0.0194	-3.24							0.0007	0.0431	-2.47	hypothetical protein	
PP3042										0.0001	0.0113	-2.79				terminase, large subunit, putative	
PP3049										<0.0001	0.0058	-2.15				conserved hypothetical protein	
PP3091				0.0028	0.0757	2.08				<0.0001	0.0032	-2.14				conserved hypothetical protein	
PP3109										0.0005	0.0187	-2.15				hypothetical protein	
PP3149										0.0006	0.0201	-2.22				transcriptional regulator, AraC family	
PP3256										0.0012	0.0308	-2.06				glycosyl transferase, group 2 family protein	
PP3316				0.0003	0.0234	-3.08										chaperone-associated ATPase, putative	
PP3376										0.0006	0.0222	-2.02				2-ketogluconate 6-phosphate reductase	kguD
PP3437				0.0003	0.0254	-3.08										CBS domain protein	
PP3466				0.0014	0.0544	2.73										ABC efflux transporter, permease/ATP-binding protein,	
PP3477										0.0003	0.0156	2.05				hypothetical protein	
PP3542										0.0001	0.0102	-2.24				conserved hypothetical protein	
PP3557										0.0010	0.0312	1.79				methyl-accepting chemotaxis transducer	
PP3595										0.0003	0.0129	-2.29				amino acid ABC transporter, permease protein	

Table 3.18 continued

locus id	cbrA::Tn5 p-value	cbrA::Tn5 FDR	cbrA::Tn5 fold-change	cbrB::Tn5 p-value	cbrB::Tn5 FDR	cbrB::Tn5 fold-change	pcnB::Tn5 locus id	pcnB::Tn5 p-value	vacB::Tn5 FDR	vacB::Tn5 fold-change	vacB::Tn5 p-value	bipA::Tn5 FDR	bipA::Tn5 fold-change	bipA::Tn5 p-value	encoded protein	gene name
PP3699												0.0003	0.0267	-2.32	hypothetical protein	
PP3702												0.0001	0.0115	-2.31	hypothetical protein	
PP3703				0.0005	0.0314	-2.91									hypothetical protein	
PP3764				0.0001	0.0091	7.35			0.0008	0.0237	-2.04				porin, putative	
PP3781															Coproporphyrinogen III oxidase fam oxygen-independent	
PP3782				0.0004	0.0262	5.10									hypothetical protein	
PP3783				0.0016	0.0577	4.70									conserved hypothetical protein	
PP3786				0.0012	0.0469	2.71									aminotransferase	
PP3846				0.0010	0.0411	-2.32									carbon-nitrogen hydrolase family protein	
PP3905				0.0001	0.0088	2.58									transcriptional regulator, LuxR family	
PP3962									0.0003	0.0128	-2.19				lipoprotein, putative	
PP4004				0.0005	0.0327	-2.03									cell division protein FtsK	ftsK
PP4017				0.0002	0.0181	2.62									conserved hypothetical protein	
PP4024												0.0003	0.0269	-2.11	ISPpu15, transposase Orf1	
PP4087												<0.0001	0.0044	-3.35	hypothetical protein	
PP4120				0.0006	0.0313	-2.42			0.0001	0.0066	2.31				NADH dehydrogenase I, B subunit	nuoB
PP4128									<0.0001	0.0056	2.59				NADH dehydrogenase I, K subunit	nouK
PP4131									0.0001	0.0105	2.23				NADH dehydrogenase I, N subunit	nuoN
PP4186									0.0001	0.0059	2.32				succinyl-CoA synthetase, beta subunit	sucC
PP4187															2-oxoglutarate dehydrogenase, lipoamide dehydrogenase	lpdG
PP4188									<0.0001	0.0017	2.22				2-oxoglutarate dehydrogenase, dihydrolipoamide succin	kgdB
PP4254									0.0001	0.0088	-2.02				hypothetical protein	
PP4272				0.0002	0.0162	-3.79						0.0001	0.0203	-2.20	conserved hypothetical protein	
PP4293									<0.0001	0.0027	-2.12				ParA family protein	
PP4334									0.0003	0.0171	2.04				flagellar motor protein MotB	motB
PP4335									0.0004	0.0219	1.84				chemotaxis protein CheZ	cheZ
PP4339									0.0030	0.0643	1.81				flagellar number regulator FleN	fleN
PP4342									0.0004	0.0171	1.99				flagellar motor switch protein FliG	fliG
PP4368									0.0008	0.0250	-1.96				flagellar basal-body M-ring protein FliF	fliF
PP4369									0.0002	0.0108	1.76				flagellar hook-basal body complex protein FliE	fliE
PP4370									0.0002	0.0124	1.93					

Table 3.18   continued

locus id	cbrA::Tn5 p-value	cbrA::Tn5 FDR	cbrA::Tn5 fold-change	cbrB::Tn5 p-value	cbrB::Tn5 FDR	cbrB::Tn5 fold-change	pcnB::Tn5 locus id	pcnB::Tn5 p-value	vacB::Tn5 FDR	vacB::Tn5 fold-change	vacB::Tn5 p-value	bipA::Tn5 FDR	bipA::Tn5 fold-change	bipA::Tn5 p-value	encoded protein	gene name
PP4382									<0.0001	0.0043	2.19				peptidoglycan hydrolase FlgJ	flgJ
PP4383									0.0026	0.0512	2.09				flagellar P-ring protein precursor FlgI	flgI
PP4386				0.0001	0.0093	-2.24			0.0033	0.0630	1.69				flagellar basal-body rod protein FlgF	flgF
PP4394				<0.0001	0.0044	-4.24									flagella basal body P-ring formation protein FlgA	flgA
PP4400				<0.0001	0.0035	-5.08									transcriptional regulator BkdR	bkdR
PP4402				<0.0001	0.0016	-7.66									2-oxoisovalerate dehydrogenase, beta subunit	bkdA2
PP4403															2-oxoisovalerate dehydrogenase, lipoamide acyltransfe	bkdB
PP4405				0.0002	0.0155	-2.89									sensory box protein	
PP4491				0.0003	0.0252	-3.02						0.0001	0.0143	-2.44	pterin-4-alpha-carbinolamine dehydratase	phhB
PP4547				0.0012	0.0474	-2.62						0.0008	0.0549	1.95	glutamine synthetase, putative	
PP4649				0.0002	0.0164	3.84									conserved hypothetical protein	
PP4650				0.0003	0.0210	4.28									ubiquinol oxidase subunit II, cyanide insensitive	cioB
PP4652				0.0001	0.0088	4.70									membrane protein, putative	
PP4697				0.0032	0.0836	-1.99									poly(A) polymerase	pcnB
PP4722									<0.0001	0.0032	2.27				transcription elongation factor GreA	greA
PP4723									0.0001	0.0077	2.34				carbamoyl-phosphate synthase, large subunit	carB
PP4725									<0.0001	0.0013	2.35				dihydrodipicolinate reductase	dapB
PP4765									<0.0001	0.0045	2.31				conserved hypothetical protein TIGR00275	
PP4783									<0.0001	0.0030	2.29				thiamine-phosphate pyrophosphorylase	thiE
PP4784									<0.0001	0.0024	2.16				glutamate-1-semialdehyde-2,1-aminomutase	hemL
PP4800									0.0003	0.0128	2.24				lipoic acid synthetase	lipA
PP4802									<0.0001	0.0028	2.50				conserved hypothetical protein	
PP4803									<0.0001	0.0027	2.67				D-alanyl-D-alanine carboxypeptidase	dacA
PP4804									0.0004	0.0147	2.16				lipoprotein, rare lipoprotein A family	
PP4813				0.0002	0.0178	-2.79			0.0003	0.0136	2.21				PAP2 family protein/DedA family protein	
PP4880				0.0009	0.0396	2.29									ribonuclease R	
PP4881				0.0002	0.0182	2.99									iron ABC transporter, periplasmic iron-binding protein	vacB
PP4903									<0.0001	0.0032	2.23				conserved hypothetical protein TIGR00157	

Table 3.18   continued

locus id	cbrA::Tn5			cbrB::Tn5			pcnB::Tn5		vacB::Tn5			bipA::Tn5			encoded protein	gene name
	p-value	FDR	fold-change	p-value	FDR	fold-change	locus id	p-value	FDR	fold-change	p-value	FDR	fold-change	p-value		
PP4905				0.0003	0.0221	-2.63			0.0002	0.0131	2.15				motility protein, MotA family	
PP4909									0.0001	0.0074	2.28				phosphoserine phosphatase	serB
PP4910									<0.0001	0.0038	2.93				conserved hypothetical protein	
PP4911									0.0002	0.0123	2.37				conserved hypothetical protein	
PP4914									0.0002	0.0136	2.27				conserved hypothetical protein	
PP4915									0.0001	0.0066	2.19				DNA topoisomerase IV, B subunit	parE
PP4929				0.0001	0.0121	-2.97						<0.0001	0.0024	-3.55	transcriptional regulator, LysR family	
PP4931												0.0002	0.0248	-2.42	conserved hypothetical protein	
PP4932				0.0018	0.0602	-2.98						<0.0001	0.0061	-3.25	conserved hypothetical protein	
PP4940									0.0008	0.0286	2.23				integral membrane protein	
PP4954									0.0001	0.0100	2.01				conserved hypothetical protein	
PP4963									0.0001	0.0083	2.01				phosphoglycerate kinase	pgk
PP4968									0.0006	0.0220	-2.64				DNA ligase, NAD-dependent, putative	
PP4977				0.0003	0.0248	3.56									5,10-methylenetetrahydrofolate reductase	metF
PP4989									0.0001	0.0068	2.17				type IV pili methyl-accepting chemotaxis transducer P	pilJ
PP4991									0.0001	0.0088	1.93				type IV pili response regulator PilH	pilH
PP4992									0.0003	0.0134	1.79				type IV pili response regulator PilG	pilG
PP5010									0.0005	0.0190	2.06				conserved hypothetical protein	
PP5011									0.0005	0.0207	2.05				ubiquinone biosynthesis methyltransferase	ubiE
PP5012									<0.0001	0.0018	2.44				conserved hypothetical protein	
PP5015									0.0002	0.0116	2.03				phosphoribosyl-ATP pyrophosphatase	hisE
PP5020									0.0001	0.0064	2.99				methyl-accepting chemotaxis transducer	
PP5025									<0.0001	0.0017	2.17				glucosyltransferase MdoH	mdoH
PP5043									0.0003	0.0151	-2.07				conserved hypothetical protein	
PP5065									0.0003	0.0133	-2.19				hypothetical protein	
PP5088									<0.0001	0.0011	-2.88				primosome assembly protein PriA	priA
PP5167									<0.0001	0.0055	-2.04				conserved hypothetical protein	
PP5184				0.0005	0.0289	-2.78									glutamine synthetase, putative	
PP5192				0.0002	0.0212	-2.58									glycine cleavage system P protein	
PP5203				0.0005	0.0290	-2.90									5-formyltetrahydrofolate cyclo-ligase family protein	gcvP-2
PP5209				0.0001	0.0112	-2.76									conserved hypothetical protein	

Table 3.18   continued

locus id	cbrA::Tn5			cbrB::Tn5			pcnB::Tn5			vacB::Tn5			bipA::Tn5			encoded protein	gene name
	p-value	FDR	fold-change	p-value	FDR	fold-change	locus id	p-value		FDR	fold-change	p-value	FDR	fold-change	p-value		
PP5213				0.0010	0.0447	-2.15										conserved hypothetical protein TIGR00148	
PP5238																conserved hypothetical protein	
PP5270	0.0003	0.0709	-2.85							0.0013	0.0338	-2.00				D-amino acid dehydrogenase, small subunit	dadA-2
PP5277				0.0014	0.0510	-2.26										major facilitator family transporter	
PP5280				0.0015	0.0537	-2.34										hypothetical protein	
PP5282										0.0001	0.0077	-2.24				ribosomal protein L28	rpmB
PP5320										0.0023	0.0523	-2.10				DNA-binding response regulator PhoB	phoB
PP5333										0.0001	0.0076	-2.18				conserved hypothetical protein	
PP5337				0.0001	0.0111	-4.29										transcriptional regulator, LysR family	
PP5398				0.0001	0.0123	-2.64										ISPpu14, transposase Orf1	

* Statistical parameters are calculated as geometric mean derived from the two biological samples for each mutant. The criteria of genes being considered to be significantly differentially expressed are p-value ≤ 0.05, FDR ≤ 0.1 and fold-change ≥ 2.0.
The box indicates the gene cluster, called *ped* cluster in *Pseudomonas putida* U (Arias et al., 2008), which is consistently repressed in all five mutants.

RESULTS AND DISCUSSION

The number of differentially regulated genes for each mutant varied strongly. For the cbrA::Tn5 and pcnB::Tn5 mutants only nine and six genes were found to be differentially expressed, respectively. A few more genes, namely 29, were detected for the bipA::Tn5 mutant whereas for the cbrB::Tn5 and vacB::Tn5 mutant significantly more genes were differentially expressed compared to the wild type, 87 and 93 genes respectively.

Interestingly, all five mutants showed a consistent down-regulation of one gene cluster that was previously described in P. putida U to be involved in the conversion of 2-phenylethanol (Arias et al., 2008).

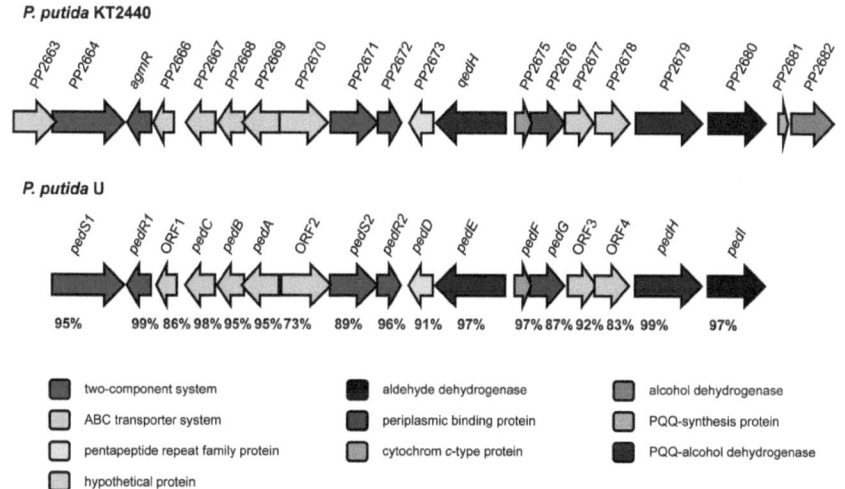

Fig. 3.28 Genetic organisation of the ped cluster in P. putida U and the homologous gene cluster in P. putida KT2440 that is required for the degradation of 2-phenylethanol. The percent values indicate amino acid similarity of encoded proteins.

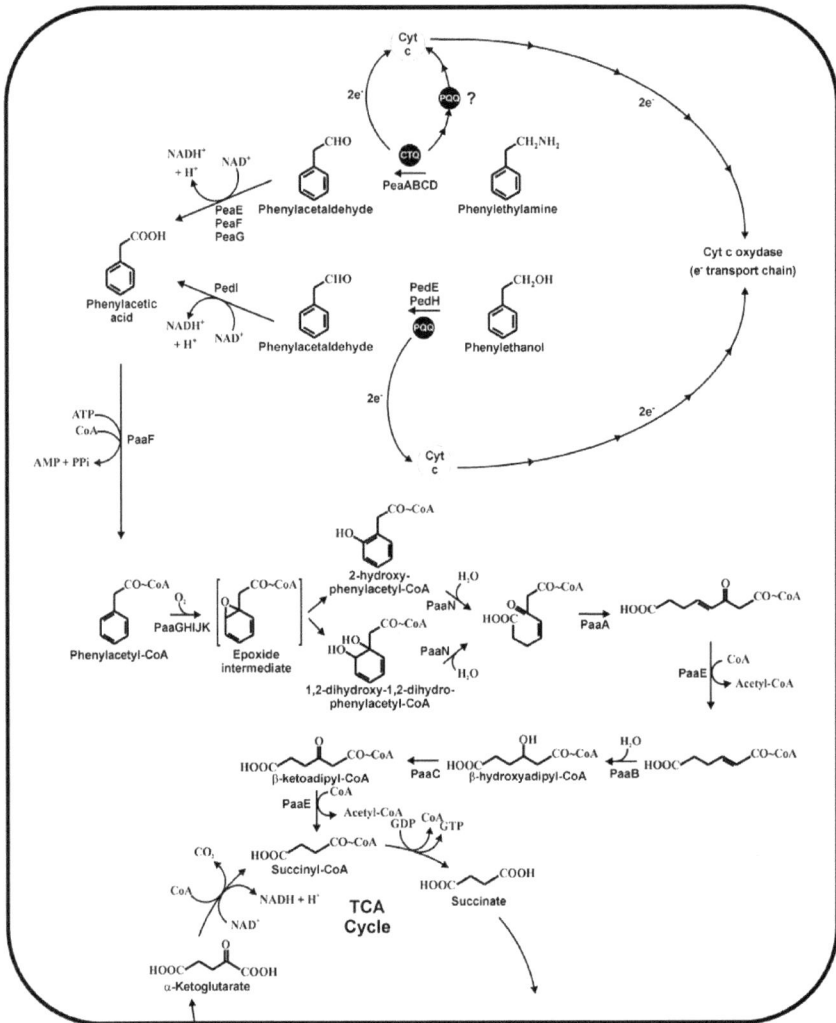

Fig. 3.29 Schematic representation of the different steps required for the aerobic transformation of 2-PhEtNH2 and 2-PhEtOH into phenylacetaldehyde in *P. putida* U. Most of these reactions seem to occur in the periplasmic space (adapted from Arias *et al.*, 2008).

RESULTS AND DISCUSSION

The Ped pathway in Pseudomonas putida

Arias *et al.* (2008) described in their study two pathways involved in the conversion of 2-phenylethylamine and 2-phenylethanol, respectively to phenylacetate via phenylactaldehyde. The gene cluster in *P. putida* U required for the uptake of 2-phenylethanol and its conversion into phenylacetate encodes two two-component systems, $PedS_1R_1$ and $PedS_2R_2$, an ABC efflux system, PedABC, a pentapeptide repeat protein, PedD, two quinoprotein dehydrogenases, PedEH, a cytochrome *c* protein, PedF, a periplasmic binding protein, PedG, an aldehyde dehydrogenase, PedI, and three hypothetical proteins. Besides the ABC efflux system, the second two-component system and the hypothetical protein named ORF2, all genes belonging to this cluster were found to be repressed upon cold shock in comparison to the wild type in at least one mutant. The homologous gene cluster in *P. putida* KT2440 consists the genes PP2664 to PP2680 with an amino acid similarity of the encoded proteins of at least 83%, while ORF2 and PP2670 share a protein similarity of 73%. Two additional gene clusters were found to be necessary for the regulation of the Ped pathway, namely PqqABCDEF and CcmABCDEFGHI, but none of the genes were found to be differentially regulated in the KT2440 mutants. These two gene clusters are involved in the biosynthesis of pyrroloquinoline quinones (PQQ) transferring electrons to a cytochrome *c* and are required for cytochrome *c* maturation respectively. The degradation of phenylacetate is then catalyzed by an additional gene cluster, the *pha* or *paa* cluster (Olivera *et al.*, 1998; Luengo *et al.*, 2001), via phenylacetyl-CoA to substrates of the carboxylic acid cycle. Neither of the genes involved in the *pha* cluster was found to be differentially expressed in the mutants that might have given a hint for the further catabolic route in the mutant background.

However, phenylacetyl-CoA can be converted into either phenylacetylglycine or phenylacetylglutamine. Though these compounds are dead-end products, they are structurally related to glycine and glutamine but differ in the additional aromatic ring. Glycine betaine and glutamate, the latter can be easily converted into glutamine, were already described to be compatible solutes, which accumulate in cells grown at low temperatures (Ko *et al.*, 1994; Park *et al.*, 1995; Brigulla *et al.*, 2003). Hence, by analogy, these compounds probably function as compatible solutes.

In a control experiment where the expression profile of the wild type and the *cbrA*::Tn*5* mutant derived from the 30°C samples were compared, the same gene cluster (PP2664-PP2680) was found to be similarly down-regulated in the mutants by about five-fold. Thus, it is likely that this gene cluster is generally down-regulated in the mutants, rather than being repressed upon cold shock. The result that the repression of this gene cluster is independent of the cold shock was furthermore supported by the data derived from the initial screening of the mutants (see 3.1.3). Though, the *vacB*::Tn5 and *bipA*::Tn5 mutants were only selected for the cold stress experiments due to their genetic background, these two mutants were also sensitive to high concentrations of benzoate. In the initial screening, all five mutants grew comparable to the wild type when cultivated in minimal medium supplemented with 15 mM benzoate, but growth was close to nil when the concentration was increased to 45 mM. Considering the consistent transcriptional profile and the sensitivity to several stress conditions of all five mutants, these data suggest that the regulation of the five genes, that are affected in the mutants background, is closely linked, and furthermore that the pathway in which the gene cluster (PP2664-2680) is involved plays an important role in stress response. In *P. putida* U a part of the *ped* cluster was described to be responsible for the conversion of 2-phenylethanol to phenylacetate; the latter is an intermediate in the phenylalanine metabolism. Comparing phenylalanine, phenylacetic acid and benzoic acid, they all have a similar structure and only differ in the benzyl side chain. Since the mutants are also benzoate sensitive, these results suggest that this gene cluster is generally involved in the conversion of aromatic amino acids and other simple aromatic compounds, and that the impairment in aromatic acid metabolism in the mutants' background leads furthermore to a cold sensitive phenotype.

In the study by Arias *et al.* (2008) transposon mutants affected in genes belonging to the *ped* cluster were tested for growth on 2-phenylethanol, benzyl alcohol or aliphatic alcohols from ethanol to decanol as sole carbon sources. The growth experiments showed that transposon mutants affected in *pedR$_1$*, *pedF*, the triple mutant in *pedABC* and the double mutant in *pedEH* could not grow on 2-phenylethanol or aliphatic alcohols from hexanol to nonanol when provided as sole carbon source. The wild type grew poorly on ethanol, hexanol and heptanol. This result showed that the *ped* cluster is also partly required for the conversion of mid-chain alcohols. For this reason, growth of the wild type and the five stress sensitive mutants was tested in minimal medium supplemented with either butanol or nonanol as sole carbon sources, on which the *P. putida* U wild type showed good and poor growth, respectively. Growth experiments were performed as in the study by Arias and co-workers: Cells were cultivated at 30°C in minimal medium (KPO$_4$H$_2$, 13.6 g/L; (NH$_4$)$_2$SO$_2$, 2.0 g/L; MgSO$_2$ x 7 H$_2$0, 0.25 g/L; FeS0$_4$ x 7 H$_2$0, 0.0005 g/L and biotine, 0.001 g/L; pH 7.0) supplemented with 0.1% butanol or nonanol, respectively. Growth was observed by measuring the optical density at 600 nm over a period of 36 hours. The growth experiments showed that *P. putida* KT2440 could not utilize nonanol; neither wild type nor mutants exhibited growth with nonanol as sole

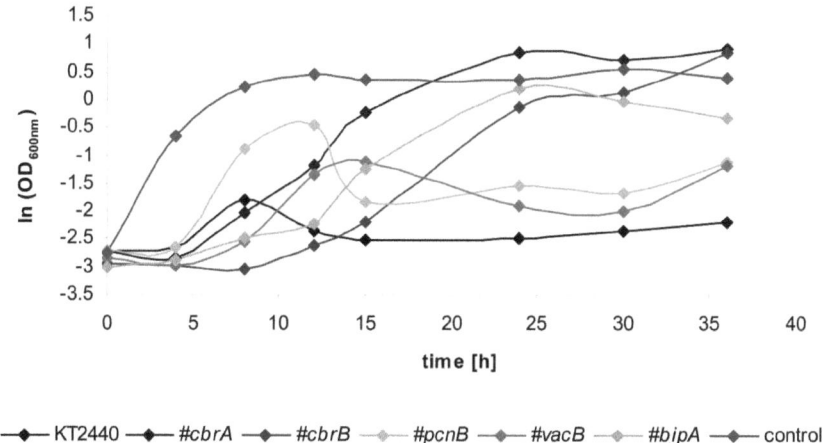

Fig. 3.30 Growth curves of *P. putida* KT2440 wild type and respective mutants in minimal medium supplemented with 0.1% butanol. As control, *P. putida* KT2440 wild type was grown in minimal medium supplemented with 15 mM succinate.

carbon source. When butanol was used as sole carbon source, growth characteristics of the strains differed from each other.

The *P. putida* KT2440 wild type grew poorly in the presence of butanol and cells started to accumulate after eight hours, similarly to the *vacB*::Tn5 culture that showed slower growth but cell aggregation after 15 hours. The *pcnB*::Tn5 culture showed initially good growth with an optical density of 0.65 after eight hours, but then cells began clumping. In contrast, the *cbrA*::Tn5, *cbrB*::Tn5 and *bipA*::Tn5 cultures grew slowly but constantly for 24 hours, when they reached stationary phase.

P. putida KT2440 is known to be solvent sensitive (Rühl *et al.*, 2009), thus it is not surprising that the strain showed only poor growth with butanol and none with nonanol as sole carbon source. Interestingly, when cultivated in minimal medium with butanol all five mutants showed better planctonic growth than the wild type, though *pcnB*::Tn5 and *vacB*::Tn5 mutants finally also formed microcolonies and aggregates. The result that the *cbrA*::Tn5, *cbrB*::Tn5 and *bipA*::Tn5 mutants showed normal growth behavior resulting in a stationary phase, and did not accumulate in the end, suggests that a knock-out of these genes rescues the sensitivity to mid-chain alcohols. Since all five mutants could initially grow better than the wild type, and all five mutants down-regulated the *ped* cluster, it is likely that this gene cluster is also involved in the conversion of aliphatic alcohols in *P. putida* KT2440 and that a repression of the *ped* cluster leads to improved growth on mid-chain alcohols.

Besides the consistent repression of the gene cluster (PP2664-PP2680) in all five mutants, the transcriptional profile of the *cbrB*::Tn5, *vacB*::Tn5 and to some degree that of the *bipA*::Tn5 mutants revealed more differentially regulated genes.

The cbrB::Tn5 transcriptome

The gene expression profile of the *cbrB*::Tn5 mutant showed 87 differentially regulated genes compared to the wild type, of which three-quarters (67) were down-regulated. Many genes encoding proteins functionally associated with cold stress were found to be differentially regulated in the *cbr*::Tn5 background, such as genes for cell wall biogenesis, cell division and RNA processing. With CbrB being a hierarchically highly ranked regulator (Nishijyo *et al.*, 2001; Zhang & Rainey, 2008; Sonnleitner *et al.*, 2009), the significant number of differentially expressed genes in response to cold shock suggests that CbrB does not only play a major role in carbon:nitrogen balancing, carbon catabolite repression and amino acid utilization, but is also important for the regulation of the cold shock response mechanisms. Indeed, out of the 87 genes that were differentially regulated in the mutant background, 17 were also identified in the transcriptome analysis of the wild type and thus are part of the key frame of cold shock response in *P. putida* KT2440. Considering the expression of the 17 genes that were found to be differentially regulated in the wild type cold shock response and again in the *cbrB*::Tn5 mutant compared to the wild type at 10°C, the *cbrB*::Tn5 mutant is affected in genes that display the major cold shock response. Sixteen out of the 17 genes were induced upon cold shock in the wild type, meaning they are required for the viability of the cells at low temperatures. In the comparison of the transcriptional profile of the wild type and the *cbrB*::Tn5 mutant, these 16 genes are down-regulated in the mutant background. The expression of these genes is comparable to the gene expression of the 30°C wild type sample, and thus the identification of these genes reveals a defect in cold shock response of the *cbrB*::Tn5 mutant.

Among these genes were seven hypothetical proteins. Though no functional prediction can be made, this result emphasizes that many unknown genes that are CbrB-dependent play an important role in cold shock response. Interestingly, the two genes PP1689 and PP1690 that were found to be highly induced upon cold shock in the wild type are CbrB-dependent, and thus the induction of this operon (together with PP1691) upon cold shock is impaired in the *cbrB*-negative background. This finding supports the hypothesis that PP1689, PP1690 and PP1691 are one of the major cold shock protein complexes in *P. putida* KT2440. Furthermore, PP2297 was found to be highly repressed in the mutant background. Integrative genetic elements play a role in DNA recombination and repair, an important mechanism in cold stress response. PP3316, a putative chaperone-associated ATPase, was also found to be impaired. This ATPase contains mainly two domains, the ClpB and AAA domain. ClpB proteins are chaperones mainly involved in protein aggregation (Lee et al., 2003); AAA ATPases are also chaperone-like proteins that assist in the assembly and disassembly of proteins (Neuwald et al., 1999). Furthermore, bkdA-2, bkdB and their transcriptional regulator bkdR were down-regulated four to seven-fold, while bkdA-2 was identified to be induced upon cold shock in the wild type background (3.6-fold with Progenika arrays and 49-fold with Illumina). BkdA-2 and bkdB are involved in valine, leucine and isoleucine degradation catabolizing in each case the second and third step in the degradation pathway. The biosynthesis and degradation of branched chain amino acids seems to be an important mechanism in the cold shock response, as many genes involved in valine, leucine and isoleucine metabolism were differentially regulated upon cold shock. Folate, in the form of its tetrahydrofolate precursors being a storage pool for DNA synthesis and S-adenosylmethionine-dependent methylation of proteins and DNA once the cells have adapted to the temperature, was already discussed in section 3.7.1. In the cbrB::Tn5 background two genes, metF and PP5203, both involved in C_1-metabolism were down-regulated of which PP5203 was identified in the transcriptional profile of the wild type. Interestingly, among the 16 genes found to be impaired in the mutant background was the two-component system PhoPQ that was already shown to be involved in membrane biogenesis (Macfarlane et al., 1999; Masschalck et al., 2003). The regulation of another gene, PP4813, encoding a PAP2 family protein, which is involved in glycerolipid metabolism was impaired in the cbrB-negative background, emphasizing the importance of cell wall maintenance upon cold shock.

These results demonstrate the essential role of CbrAB as one of the major regulators, since many important genes in the cold shock response are CbrB-dependent. Most of the genes, 67 out of 87, are activated by CbrB as indicated by the repression in a cbrB-negative background, whereas the remaining genes are repressed by CbrB. The important role of CbrAB in cold shock response is even more emphasized as 25% (16 out of 67) of the impaired genes in the cbrB::Tn5 mutant belong to the key players in cold shock response in P. putida KT2440 as these were identified by all three transcriptome platforms. The differential regulation of these 16 genes displays the defect of the cbrB::Tn5 mutant in cold shock response, since the expression is similar to that of the 30°C wild type samples, and thus leads to the cold sensitive phenotype.

The CbrB-dependent expression of these genes was supported by the fact that pcnB was repressed. PcnB expression in P. fluorescens was previously reported to be CbrB-dependent (Zhang et al., 2010). In a ß-galactosidase activity assay they monitored the expression of pcnB in wild type and in a cbrB-negative mutant. In the mutant background they could hardly detect any pcnB expression, indicating the regulatory role of CbrB. By the identification of CbrB-dependent expression of *pcnB*, the data implicated that CbrAB might link carbon metabolism and mRNA degradation. This hypothesis was furthermore supported by the data of this study as *vacB* being involved in RNA degradation was also found in the mutant background, though CbrB seems to repress *vacB* as it was induced in the *cbrB*::Tn5 mutant.

The vacB::Tn5 transcriptome

VacB is known to be involved in several RNA processes such as stability, quality control and decay (Cheng & Deutscher, 2005; Andrade *et al.*, 2006), but was also described to be important for host cell colonization of an endosymbiontic *Burkholderia* strain (Ruiz-Lozano & Bonfante, 2000) and for virulence in *Shigella flexneri*, *E. coli* (Tobe *et al.*, 1992; Cheng *et al.*, 1998) and *Brucella abortus* (Miyoshi *et al.*, 2007).

The transcriptional profile of the *vacB*::Tn5 mutant showed only minor differences in regulation compared to the wild type; the absolute fold-changes were between 1.7 and 3.9. The reason for considering genes to be significantly differentially regulated though the absolute fold-change was below 2.0, was that these genes were located within operons that were differentially regulated or that homologous genes encode a functionally related protein. In this case, many genes encoding methyl-accepting chemotaxis transducers were found. In the *vacB*-negative background, two processes seemed to be mainly affected. These were cell wall biogenesis and motility; with the former being indicated by the differential regulation of glycosyl transferases and elements of the peptidoglycan layer.

Studies of *vacB* in *Aeromonas hydrophila* (Erova *et al.*, 2008) and *Helicobacter pylori* (Tsao *et al.*, 2009) furthermore demonstrated that the exoribonuclease *vacB* is involved in motility. In pathogenic bacteria, motility involving flagella is one important factor contributing to virulence (Ottemann & Lowenthal, 2002). In *Aeromonas hydrophila*, an isogenic mutant lacking *vacB* showed reduced motility but was furthermore unable to grow at 4°C. At standard temperature, 37°C, the mutant grew comparable to the wild type (Erova *et al.*, 2008). The temperature sensitivity was supported by a microarray study of *Helicobacter pylori* wild type compared to a *vacB*-negative mutant (Tsao *et al.*, 2009). Besides that the transcriptional profile of the *vacB*-negative mutant displayed six-virulence related genes associated with motility, chemotaxis and apoptosis that were posttranscriptionally repressed by *vacB*, the exposure to different environmental conditions such as pH and temperature also caused an alteration of gene expression in the *vacB*-negative mutant.

A high number of the 93 differentially expressed genes in the *vacB*-negative background are involved in chemotaxis and motility. Six chemotactic transducers and *cheZ* encoding a chemotaxis protein were identified. These transducer proteins are transmembrane methyl-accepting proteins that are required for chemotactic behaviour of flagellated, motile bacteria (Springer *et al.*, 1979; Hazelbauer & Harayame, 1983). *MotB*, *fliEFG*, *flgFIJ* and PP4905 encoding a motility protein of the MotA family were all differentially expressed and are involved in flagella assembly. Three genes involved in signal transduction related to type IV pili, *pilGHJ*, were induced compared to the wild type.

Since many differentially regulated genes in the *vacB*-negative background are involved in chemotaxis and motility, the motility of the *vacB*::Tn5 mutant was tested in a motility assay including swimming, swarming and twitching (see 2.5.8). The *P. putida* KT2440 wild type, the *vacB*::Tn5 mutant and as control the *Pseudomonas aeruginosa* PA14 strain were incubated overnight at 30°C on the respective plates.

Table 3.19 Motility assay. Swarming, swimming and twitching motility was monitored by measuring the colony diameters (cm)*.

	swarming	swimming	twitching
KT2440	0.5	2.0	0.55
***vacB*::Tn5**	0.5	1.6	0.7
PA14	0.9	2.35	1.8

* The colony diameter is given as mean of two experiments. The value in brackets is the colony diameter after two days of incubation.

The motility assay showed that the *vacB*::Tn5 mutant is slightly affected in swimming motility compared to the *P. putida* KT2440 wild type. The swimming motility is enabled by flagella (Murray & Kazmierczak, 2006). The defect of swimming motility of the *vacB*::Tn5 mutant was reflected by the induction of representative genes of the *fli* and *flg* gene clusters that are involved in flagellar assembly in order to compensate for the limited swimming capability. In contrast to swimming and swarming, twitching motility is flagellar-independent and enabled by type IV pili (Mattick, 2002). Apart from *pilGHJ* genes, which are related to type IV pili, no significant differences in twitching motility were observed between the wild type and the *vacB*::Tn5 mutants.

The studies on *Aeromonas hydrophila* and *Helicobacter pylori* strongly support the transcriptional data derived from the *vacB*-negative mutant in *P. putida* KT2440 as most of the genes found to be differentially regulated compared to the wild type are involved in chemotaxis and motility. Furthermore, the phenotypic assays demonstrated the strong cold sensitivity of the *vacB*::Tn5

RESULTS AND DISCUSSION

mutant as the growth rate was strongly reduced after cold shock compared to the wild type (see 3.2.2). This suggests that *vacB* plays a more global role in sensing environmental changes and in regulation stress associated mechanisms.

In contrast to the transcriptional profile of the *cbrB*::Tn5 mutant where 16 genes were found to display a defect in cold shock response, only a few genes were found in the *vacB*-negative background that displayed a similar expression as the 30°C wild type samples. Three genes were found to be impaired in the *vacB*-negative background, since these genes displayed a similar expression to the 30°C wild type sample. Interstingly, all three genes are related to energy conversion; *sucC* and *ipdG* are involved in the tricarboxylic acid cycle and *nuoN* in oxidative phosphorylation. The energy metabolism was generally down-regulated in the wild type upon cold shock. That genes representative for energy metabolism are up-regulated in the *vacB*-negative background, suggests that the *vacB*::Tn5 mutant is slightly affected in energy conversion, since the regulation of the respective genes is similar to the wild type at 30°C. Three additional genes (*kgdB*, *nuoB* and *nuoK*) related to either the tricarboxylic acid cycle or oxidative phosphorylation were found to be induced in the *vacB*-negative background. Though these genes do not directly overlap with the transcriptome data of the wild type, this finding supports the hypothesis that the *vacB*::Tn5 mutant is affected in energy conversion, since these genes are representative for the same regulatory networks.

The bipA::Tn5 transcriptome

In the *bipA*::Tn5 mutant 22 genes were differentially regulated besides genes belonging to the *ped* cluster, among which 12 hypothetical proteins and three transcriptional regulators were found. Besides these genes, PP1116, a site-specific recombinase, PP1248 encoding a transporter related protein and PP4024, a transposase, were found to be differentially regulated in the *bipA*-negative background. Thus, a role for BipA in cold stress response is supported, since these genes exhibit cold stress associated functions such as DNA recombination and repair (PP1116, PP4024) and membrane biogenesis (PP1248) as discussed previously. Out of these genes, only one gene was found to display a defect in cold shock response. The LsyR family transcriptional regulator, PP4929, was induced upon cold shock in the wild type, according to the Progenika array results by 3.3-fold (not listed in Table 3.16), and was also differentially regulated in the *cbrB*::Tn5 mutant. PP4929 is predicted to be organized in an operon together with *kdtA*, which is involved in polysaccharide biosynthesis. The expression of this regulator in the two mutants was comparable to the expression of the wild type at 30°C, indicating that the transcriptional regulator PP4929 plays an important role in cold shock response, since the two mutants are impaired in its regulation upon cold shock and its functionality is related to cold stress mechanisms.

BipA has already been hypothesized to be a novel regulatory protein by directly interacting with the ribosomes and to be responsible for the efficient translation of target genes (Krishnan & Flower, 2008). Furthermore, it was shown that Fis expression, which is a highly ranked transcriptional modulator in *E. coli*, is BipA-dependent (Owens *et al.*, 2004). The identification of three transcriptional regulators whose expressions seem to be BipA-dependent support the results of the previous studies, that BipA is an important regulatory protein involved in stress response.

The transcriptome data of the cold sensitive mutants revealed three major aspects. Firstly, all five transposon mutants are affected in the same gene cluster (PP2664 to PP2680), called the *ped* cluster in *P. putida* U (Arias *et al.*, 2008). Though the repression of this gene cluster was not caused by the cold shock as such, it is likely that this pathway plays an important role in cold shock response, because all five mutants show a consistent cold sensitive phenotype Secondly, the transcriptome profile of the *cbrB*::Tn5 mutant revealed many genes that are CbrB-dependent. Many of the identified key players in cold shock response of *P. putida* KT2440 are induced by CbrB, emphasizing once more the high rank of the CbrAB two-component system within the regulatory hierarchy in *P. putida* KT2440 physiology. Thirdly, *vacB* was demonstrated to play a more global role in *P. putida* than only being involved in RNA processing. In fact, it is an important factor for the response to a temperature decrease as displayed by the cold sensitive phenotype and processes being involved in cold shock response such as cell wall maintenance were affected on the transcriptional level in a *vacB*-negative background. This was already supported by the proteome results as the proteome profile showed an increase of VacB as well as BipA upon cold shock.

RESULTS AND DISCUSSION

In summary, the highly ranked regulatory function of CbrB was demonstrated by the transcriptional data of the mutants, since only the transcriptional profile of the *cbrB*::Tn5 mutant displayed a direct defect in cold shock response. In total, 16 genes showed an expression comparable to the wild type at 30°C, which belong to the key players of cold shock response in *P. putida* KT2440. VacB was previously described to be involved in chemotaxis and motility (Erova *et al.*, 2008, Tsao *et al.*, 2009), thus it is likely that the large number of differentially regulated genes that are associated with chemotaxis and flagellar assembly reflect the general phenotype of the *vacB*::Tn5 mutant, and that these genes are not differentially regulated in response to cold shock, since they were not found to be differentially regulated in the cold shock response of the wild type.

The transcriptional profiles of the five stress sensitive mutants revealed the consistent repression of one gene cluster (PP2664-PP2680). The results by Arias and co-workers showed that the *ped* cluster is required for the conversion of 2-phenylethanol to phenylacetate, an intermediate in the phenylalanine catabolism. This study revealed that this particular gene cluster is consistently repressed in all five stress sensitive mutants that are cold, but also benzoate sensitive. Furthermore, it was shown that the *bkd* operon is affected in the *cbrB*-negative background, and Madhusudhan and co-workers (1993, 1997) reported previously that BkdR regulation which is the transcriptional activator of the *bkd* operon is VacB-dependent. The *bkd* operon is involved in the degradation of the branched chain amino acids valine, leucine and isoleucine. The metabolism of branched chain and aromatic amino acids seemed to be an important mechanism in cold shock response, as five genes (*leuC*, *ilvCD* and *ilvBN*), directly involved in either valine or isoleucine biosynthesis and one gene in tryptophan metabolism (*trpB*) were found to be repressed in the wild type upon cold shock. The study by Zhang *et al.* (2010) demonstrated the CbrAB-dependent regulation of *pcnB*, which was supported by the transcriptome data derived from the *cbrB*::Tn5 mutant in this study. In a *cbrB*-negative background, the expression of *pcnB* was repressed. The deletion of *pcnB* resulted in a reduced growth rate under various growth conditions, suggesting that polyadenylation is important for the adaptation processes to changing environmental conditions. Furthermore, the expression of *vacB* seems to also be CbrB-dependent with CbrB being a negative regulator, since *vacB* expression was induced (2.3-fold) in the *cbrB*::Tn5 mutant. Furthermore, the transcriptome data of the mutants showed that the regulation of the transcriptional regulator PP4929, which is involved in polysaccharide biosynthesis, is an important mechanism in cold shock response, since its regulation was found to be affected in the *cbrB*::Tn5 and *bipA*::Tn5 mutants. This supported firstly the defect in cold shock response of these mutants and secondly showed that the CbrB- or BipA-dependent regulation of target genes is partially overlapping.

These results support that the regulation of these five genes is closely linked with CbrAB and PcnB being a regulatory unit, the regulation of VacB being CbrB-dependent, and CbrB- and BipA-dependent regulation partially targeting the same genes. VacB and BipA were already described to be required for growth at low temperatures (Pfennig & Flower, 2001, Cairrão *et al.*, 2003, Reva *et al.*, 2006). Although none of the five genes was identified to be affected upon cold shock on the transcriptional level in the wild type, VacB and BipA were found with the proteomics approach, where both proteins were induced in response to decreasing temperatures. In addition, CbrAB was already shown to regulate the expression of the small RNA CrcZ, which is the global regulator of carbon catabolite repression. CrcZ sequesters Crc, which is a RNA-binding protein, and thus enables the expression of Crc target genes (Sonnleitner *et al.*, 2009). Crc and exonucleases were already shown to share some sequence similarity (MacGregor *et al.*, 1996), and furthermore to exhibit overlapping regulatory functions. Crc and VacB are both involved in *bkdR* expression (Madhusudhan *et al.*, 1993, 1997). These data provide a direct link between central metabolism, transcription and translation efficiency. CbrAB is well characterized to be required for a healthy carbon:nitrogen balance in the cell. By regulating the expression of *crcZ*, CbrB is directly involved in carbon catabolite repression and by thus the transcription of many genes is CbrB-dependent such as *benR*, *amiE* (Sonnleitner *et al.*, 2009) and *bkdR* (Hester *et al.*, 2000). The CbrB-dependent regulation of *pcnB* and *vacB* combines the regulation of the central metabolism with RNA quality control, since the two genes are involved in mRNA degradation as well as in mRNA stability (Sarkar, 1997, Andrade *et al.*, 2006). Considering the CbrB-dependent regulation of *benR* which is involved in benzoate degradation, *amiE* (though there is no homologue in *P. putida* KT2440) in the conversion to phenylacetate or benzoate and *bkdR* in the metabolism of branched chain amino acids, the transcriptome and metabolome data strongly suggest that the metabolism of aromatic amino acids and other related compounds is highly affected upon cold shock. This was displayed by the accumulation of amino acids such as tyrosine and valine, but as well by the benzoate sensitivity of the five mutants. The regulation of complex amino acids seems to be affected by either one of the five knock-outs leading to the cold sensitive phenotype.

3.9 Metabolome Profile of *Pseudomonas putida* KT2440 Wild Type and Stress Sensitive Tn5 Mutants

Metabolome analysis is, in addition to transcriptome and proteome studies, the third field in functional genomics, and enables the identification of new metabolic pathways. It facilitates the investigation of the metabolome profile under changing growth conditions and thereby helps to elucidate adaptation processes to stress conditions such as non-preferred carbon sources, pH and temperature on the intra- and extracellular metabolite level (van der Werf *et al.*, 2008, Frimmersdorf *et al.*, 2010).

In this study, the changes of intracellular metabolite composition of *P. putida* KT2440 wild type and the respective stress sensitive Tn5 mutants to a sudden decrease in temperature was investigated. The metabolome experiments were conducted in collaboration with Christian Jäger from the Department of Bioinformatics and Biochemistry at the Technical University of Braunschweig. For this, aliquots of bacterial suspension from the same cultures as for transcriptome analysis were used for sample preparation (see chapter 2.9). Metabolite extraction was performed in Hannover according to the provided protocol, and samples were sent to Braunschweig for final sample processing and data acquisition. In total six technical replicates were processed in parallel for each condition (30°C and 10°C) and the experiments were repeated twice for each strain (wild type and five Tn5 mutants).

The metabolite analysis is GC-MS-based. According to the retention times and mass spectra, about 507 different compounds were found in at least one sample. All peaks with a peak area less than six were excluded from the analysis, because this value was defined as the threshold for background noise in this study. The total peak area of the excluded compounds accounted for less than 1% of the total peak area of all compounds detected. Out of these 507 compounds, 105 could be identified as known metabolites by comparison with the in-house metabolite libraries in Braunschweig. The remaining 402 compounds could be identified as distinct compounds, but the representative metabolite is unknown so far. Thus, these compounds were declared as non-matches, and are not included in the further analysis of the metabolome profile of the strains. In the following (metabolite) concentration instead of peak area will be used to ease readability. Though metabolite concentration and peak area is compound-specific, meaning concentration of a specific metabolite within one sample is not directly correlated to the peak area, the term concentration is here feasible, since the very same metabolites will be compared among different conditions and strains.

Data normalization

During sample preparation ribitol is supplied to each sample at a defined concentration (2 g/L) as external standard for data normalization. In total, two biological replicates of six (wild type and five Tn5 mutants) independent experiments were processed. Each experiment included two conditions (30°C and 10°C), and each condition was represented by six technical replicates. There were 144 samples, of which one from the *bipA*::Tn5 10°C (experiment 1) could not be analyzed due to technical problems during GC-MS measurement. In this study, data were normalized as follows: The total peak area of all metabolites per sample (per technical replicate) was divided by the peak area of ribitol, which was applied as standard during sample preparation. This resulted in a normalization factor per sample. Since the 12 samples of one condition (e.g. wt 10°C experiment 1 and 2) are supposed to represent the same metabolic profile with equal concentrations of respective metabolites, the median normalization factor per condition was calculated. The normalization factor per sample was divided by the mean normalization factor per condition. This resulted in the correction factor, which was used for the normalization of the peak areas. Thereafter, the median of the peak area per condition was calculated. A list with the median peak areas per condition and strain is provided in Table 3.20 and 3.21. Only known metabolites which were found in the majority of samples are listed.

For visualization and interpretation, the fold-change per metabolite and condition (combining experiment 1 and 2) was calculated by referring to the median peak area over all 24 conditions (2 biological replicates, six experiments, 2 conditions). In Figure 3.31, the fold-change per condition within one metabolite is given in numbers representing the absolute fold-change compared to the median peak area over all conditions, and is furthermore presented in a red and blue color-code for visualization.

RESULTS AND DISCUSSION

Table 3.20 Metabolite concentrations provided as median peak area* of six technical replicates for *Pseudomonas putida* KT2440 wild type.

metabolite	Number of samples found[a]	Number of samples found per condition[b]	experiment 1 wt 10°C	experiment 1 wt 30°C	experiment 2 wt 10°C	experiment 2 wt 30°C
1-monooleoylglycerol	65	3336	120.8	438.6	438.3	2120.4
1-monopalmitoylglycerol	102	5656	1613.0	2252.9	1828.4	3752.0
1-pyrroline-3-hydroxy-5-carboxylate	143	6666	107808.7	86314.7	118602.4	41906.5
2-keto-3-deoxygluconic acid	93	6666	15979.4	20090.9	22414.0	35049.6
2-oxoglutaric acid	93	6566	47677.4	11778.2	22571.5	24523.7
3-hydroxybutyric acid	75	6666	9069.4	5023.6	25221.1	7726.5
3-phosphoglyceric acid	94	6666	392.5	209.0	390.7	468.6
5'-AMP	120	6566	1387.0	6886.5	3291.9	8993.5
alanine	142	6666	215224.2	66662.5	165505.7	67004.7
aspartic acid	143	6666	143004.7	141008.5	192967.8	82038.1
beta-alanine	143	6666	30945.8	85741.9	35354.1	128732.2
citric acid	116	6666	18177.9	1443.1	2870.4	2845.1
dodecanoic acid	125	2243	100.0	100.0	15167.0	2284.9
fructose 6-phosphate	136	6666	355258.6	59130.4	306269.4	60029.3
glucosamine 6-phosphate	98	6666	19997.6	5329.5	16882.4	5327.3
glucose	141	6666	21198.0	20224.2	24333.7	24708.3
glucose 6-phosphate	143	6666	498103.6	126662.5	559546.6	139997.1
glutamic acid	143	6666	2345420.5	1412956.1	2717664.4	1246619.6
glutamine	135	6563	47753.6	50992.1	60793.5	609.9
glycerol	142	6666	3905.3	4228.6	8568.6	6261.9
glycerol 3-phosphate	143	6666	33766.9	36460.6	68827.7	35654.7
glycine	143	6666	55773.7	47338.7	51584.4	36684.0
homoserine	101	0000	100.0	100.0	100.0	100.0
hydroquinone	83	0000	100.0	100.0	100.0	100.0
inosine	119	6655	2962.0	5365.3	4336.8	7582.1
isoleucine	72	1100	100.0	100.0	100.0	100.0
lactic acid	121	6666	321656.2	91655.8	179341.3	136196.3
lactose	88	6665	5092.2	7998.1	9858.3	11505.4
lysine	143	6666	24740.9	36643.6	24233.5	34860.5
malic acid	98	6655	7922.4	320.5	412.3	309.8
maltose	98	6666	1993.2	28163.7	5329.1	16473.5
mannose 6-phosphate	117	6666	87302.8	23372.2	101995.1	32143.4
N-acetylglutamic acid	143	6666	61963.9	46755.7	69345.4	50074.6
nicotinamide	128	6666	11511.6	15980.5	14478.4	23919.9
ornithine	104	2464	100.0	6771.3	11049.4	13106.5
oxalic acid	134	6666	38049.8	42211.8	62634.4	54461.0
palmitic acid	143	6666	81867.4	107876.9	88922.6	139485.6
phosphoethanolamine	106	3356	545.9	539.0	1706.7	1553.0
phosphoric acid	77	1001	100.0	100.0	100.0	100.0
putrescine	143	6666	104562.7	106999.0	114100.0	179543.2
pyrophosphoric acid	86	6445	9232.5	13630.8	8286.1	13432.8
serine	135	6604	10093.2	16856.7	100.0	11236.3
spermidine	86	6263	4415.9	100.0	3711.9	489.3
succinic acid	143	6666	71069.5	34006.2	27631.1	26594.8
tartaric acid	79	0000	100.0	100.0	100.0	100.0
threonine	143	6666	64349.9	42397.3	56244.6	35197.4
triethanolamine	77	0001	100.0	100.0	100.0	100.0
tyrosine	69	3141	2795.7	100.0	6521.0	100.0
uracil	101	4353	2110.0	2072.5	2320.6	662.7
valine	143	6666	1163885.0	120082.1	1284037.4	92399.0
xylulose 5-phosphate	66	6665	6675.3	6690.6	6690.9	7501.9

The relative error was calculated for metabolites, which have been detected in at least four out of six technical replicates per condition. On average, 22% relative error were calculated for all metabolites of one condition.

[a] Total number of samples, in which the respective metabolite was detected (in total 143 samples were analyzed).

[b] Number of samples per strain and condition, in which the metabolite was detected (6 technical replicates per condition, in order 10°C, 30°C of experiment 1; 10°C, 30°C of experiment 2). For experiment 1 only 5 samples of the 10°C condition were analyzed.

* The median peak area was calculated from the raw data provided by the GC-MS analysis. 100.0 means the metabolite was not detected.

RESULTS AND DISCUSSION

Metabolite	KT2440 wt 10°C	KT2440 wt 30°C	cbrA::Tn5 10°C	cbrA::Tn5 30°C	cbrB::Tn5 10°C	cbrB::Tn5 30°C	pcnB::Tn5 10°C	pcnB::Tn5 30°C	vacB::Tn5 10°C	vacB::Tn5 30°C	bipA::Tn5 10°C	bipA::Tn5 30°C
1-monooleoylglycerol	0.9	4.8	1.3	2.0	3.9	0.4	5.6	1.4	7.9	6.3	9.3	0.4
1-monopalmitoylglycerol	0.8	1.4	0.3	1.1	1.0	0.5	1.2	0.9	2.0	2.0	2.4	1.4
1-pyrroline-3-hydroxy-5-carboxylate	1.9	1.1	1.8	0.8	0.6	0.6	2.3	0.6	1.8	0.7	1.7	0.6
2-keto-3-deoxygluconic acid	1.8	2.6	2.2	2.3	0.0	0.8	0.4	0.1	2.9	2.4	0.8	0.0
2-oxoglutaric acid	6.1	3.2	6.0	1.6	0.7	0.0	0.5	0.1	1.5	0.9	2.9	0.3
3-hydroxybutyric acid	10.7	4.0	0.1	0.3	0.1	0.1	9.6	0.8	27.2	4.2	3.0	0.1
3-phosphoglyceric acid	1.8	1.5	2.7	2.5	0.4	1.3	1.7	0.8	1.0	0.8	1.4	0.6
5'-AMP	2.2	7.5	3.2	4.6	0.8	1.3	0.7	1.0	0.1	0.4	1.9	0.6
alanine	1.3	0.4	1.3	1.6	0.3	0.8	2.1	0.7	2.4	0.8	2.2	0.6
aspartic acid	1.4	0.9	2.0	0.7	0.8	0.4	1.8	0.5	1.2	0.8	2.7	0.6
beta-alanine	0.9	3.1	0.7	3.6	0.3	1.6	0.6	0.8	1.4	1.8	0.8	1.2
citric acid	6.0	1.2	3.1	0.7	0.7	0.3	0.8	0.1	1.8	1.0	1.6	0.8
dodecanoic acid	0.9	0.1	2.4	4.1	0.8	2.0	0.5	0.7	3.9	2.9	1.5	1.2
fructose 6-phosphate	13.6	2.4	7.5	0.8	0.4	0.1	2.6	0.1	11.5	0.6	11.1	0.4
glucosamine 6-phosphate	4.5	1.3	7.0	0.1	0.2	0.0	13.0	0.4	3.8	0.0	4.6	0.3
glucose	1.1	1.1	2.5	2.3	0.3	0.5	3.9	2.4	2.3	1.1	0.9	0.3
glucose 6-phosphate	4.3	1.1	4.9	0.6	0.4	0.1	2.2	0.1	5.0	0.3	3.9	0.3
glutamic acid	1.9	1.0	2.1	1.1	1.4	1.0	1.3	0.4	1.7	0.6	1.2	0.5
glutamine	2.0	1.0	2.4	0.6	0.5	0.2	2.7	0.4	3.8	0.8	3.5	0.2
glycerol	1.4	1.2	2.0	0.9	0.7	0.6	0.9	0.5	1.3	1.0	3.6	7.4
glycerol 3-phosphate	2.9	2.0	1.4	1.0	0.6	0.6	1.3	0.7	1.6	0.7	1.1	0.4
glycine	1.7	1.3	1.2	0.8	0.4	0.4	0.9	2.3	1.6	1.0	1.7	0.6
homoserine	0.5	0.0	1.5	2.4	1.2	1.7	1.2	3.5	2.5	1.8	1.6
hydroquinone	0.1	0.1	1.8	1.2	0.4	0.4	2.3	1.9	1.5	1.3	2.1	2.1
inosine	1.1	2.0	0.8	1.4	0.1	0.6	1.9	1.7	0.9	1.1	1.5	2.0
isoleucine	0.0	0.0	2.1	4.8	0.5	3.9	5.0	1.4	13.9	1.6	5.4	1.7
lactic acid	5.5	2.5	3.2	2.8	0.7	0.6	1.9	0.6	1.6	0.2	1.8	0.2
lactose	1.7	2.2	1.6	1.9	2.0	5.7	0.6	0.0	1.0	0.6	0.3	0.2
lysine	0.7	1.0	0.6	1.0	0.1	0.6	1.4	0.9	2.9	1.1	1.5	0.8
malic acid	16.1	1.2	1.0	1.7	0.4	0.3	1.5	0.8	1.3	0.8	1.4	0.6
maltose	0.8	4.9	1.9	0.1	36.8	0.0	0.0	1.2	1.1	0.0	0.0
mannose 6-phosphate	6.0	1.6	6.5	0.8	0.4	0.1	1.5	0.0	8.4	0.5	4.6	0.3
N-acetylglutamic acid	2.2	1.6	1.7	0.5	0.3	0.1	1.2	0.6	1.1	1.5	1.6	0.7
nicotinamide	0.7	1.1	1.1	1.6	0.2	0.8	1.0	0.8	1.8	1.4	1.0	1.1
ornithine	0.9	1.7	2.7	1.3	1.2	0.1	0.9	0.5	1.7	1.2	0.9	0.0
oxalic acid	1.3	1.2	0.7	2.6	0.6	0.7	0.5	0.4	3.6	3.0	1.5	1.2
palmitic acid	0.7	1.0	0.7	1.5	0.7	0.9	1.0	1.1	1.7	1.7	1.7	1.6
phosphoethanolamine	0.4	0.3	1.2	1.4	2.3	2.1	0.8	0.2	4.8	1.6	4.7	0.5
phosphoric acid	0.0	0.0	1.2	0.6	2.0	2.0	0.9	1.8	1.5	1.2	1.0	1.2
putrescine	1.0	1.3	0.9	0.7	0.4	0.9	0.9	0.7	1.8	1.5	0.8	0.7
pyrophosphoric acid	1.3	2.0	2.1	2.8	3.0	1.1	0.9	0.2	0.9	1.3	0.5	1.0
serine	0.4	1.2	0.8	1.3	0.3	0.8	0.7	1.3	1.3	1.6	1.5	1.4
spermidine	3.2	0.2	2.9	1.2	1.0	0.8	2.6	1.1	0.1	5.1	0.6	0.2
succinic acid	3.2	1.9	1.2	1.1	0.4	1.3	0.6	1.8	1.1	1.5	0.6
tartaric acid	0.0	0.0	1.0	3.9	0.3	0.6	1.6	2.3	2.0	3.2	3.2	0.6
threonine	1.5	0.9	1.3	1.1	0.4	0.7	1.6	0.5	1.9	0.8	1.9	0.8
triethanolamine	0.1	0.1	2.7	3.3	1.6	2.9	0.1	0.2	3.7	3.9	2.8	1.2
tyrosine	4.4	0.1	0.1	4.2	0.1	1.0	5.7	2.2	17.9	0.5	8.4	3.1
uracil	0.8	0.5	3.1	2.3	3.0	1.3	2.5	3.4	1.4	4.0	4.5	4.2
valine	5.0	0.4	3.0	0.7	0.1	0.3	4.2	0.5	5.0	1.1	3.7	0.5
xylulose 5-phosphate	13.4	14.2	3.1	3.8	0.2	1.2	0.5	0.2	0.2	2.6	11.3	0.7
2-hydroxyglutaric acid	23.9	20.0	1.0	1.0	1.0	1.0	1.0	1.0	14.6	1.0	1.0	1.0
4-aminobutanoic acid	79.6	71.6	1.0	1.0	9.1	1.0	1.0	1.0	1.0	1.0	1.0	1.0
5-aminolevulinic acid	12.5	16.9	1.0	1.0	1.0	1.0	1.0	1.0	1.0	1.0	1.0	1.0
5-oxoproline	208.4	118.7	1.0	1.0	1.0	1.0	1.0	1.0	114.2	1.0	1.0	1.0
benzoic acid	340.8	377.6	1.0	40.8	1.0	1.0	1.0	92.8	1.0	269.5	1.0	44.0
cellobiose	2.2	19.9	1.0	1.0	1.0	1.0	1.0	1.0	1.0	1.0	1.0	1.0
cytosine	1.0	22.9	1.0	1.0	1.0	1.0	1.0	1.0	63.8	1.0	118.3	58.1
ethanolamine	11.0	133.6	1.0	6.8	1.0	1.0	1.0	1.0	6.5	168.9	23.8	1.0
fructose	9.1	1.0	24.0	1.0	1.0	1.0	1.0	6.1	47.6	19.8	35.7	2.9
fumaric acid	94.5	2.9	1.0	1.0	1.0	1.0	1.0	1.0	10.9	1.0	1.0	1.0
hydroxylamine	1.0	12.0	1.0	14.4	1.0	14.5	1.0	10.0	1.0	1.0	1.0	14.5
N-acetylglucosamine	6.9	54.6	1.0	1.0	1.0	1.0	1.0	1.0	1.0	1.0	1.0	1.0
phenylalanine	19.4	1.0	1.0	1.0	1.0	1.0	59.0	1.6	1.0	147.5	335.2	34.6
proline	1.0	1.0	621.3	1.0	1.0	1.0	23.8	421.1	14.5	73.4	45.0	726.8
pyruvic acid	1085.3	2099.2	1.0	42.9	1.0	79.0	1.0	24.0	1.0	1.0	1.0	1.0
rhamnose	12.0	5.7	13.6	1.0	1.0	1.0	10.8	1.0	24.0	17.9	1.0	1.0
ribose	125.2	171.7	259.0	1.0	89.8	1.0	1.0	1.0	1.6	18.6	1.0	1.0
ribulose 5-phosphate	96.0	154.2	20.6	1.0	1.0	1.0	1.0	1.0	9.6	1.0	40.9	35.6
saccharopine	10.8	9.8	1.0	1.0	1.0	1.0	1.0	1.0	1.0	1.0	1.0	1.0
tryptamine	15.1	1.0	1.0	1.0	1.0	1.0	13.8	1.0	96.3	1.0	1.0	1.0
tryptophan	1.0	1.0	1.0	1.0	1.0	1.0	1.0	1.0	37.1	58.4	62.0	1.0

Fig. 3.31 The graph displays the fold-change of concentration per condition compared to the median concentration of the respective metabolite of all conditions. Blue and red colors indicate an over- (red) or under-representation (blue) of the respective metabolites per condition. The upper block contains metabolites that were found in the majority of samples, whereas the bottom block contains selected metabolites that were only detected in certain conditions.

A general analysis of the metabolite concentrations of the 30°C and 10°C samples indicated that intermediary metabolism is reduced at 10°C. The percentage peak area of known metabolites was consistently higher at 10°C compared to that of the 30°C samples. At 30°C, more unknown compounds were consistently identified, reflecting a more complex constitution of metabolites within the cells. This finding correlates very well with the transcriptome data, since many genes involved in intermediary metabolism were repressed upon cold shock (see Figure 3.24). Out of the 105 metabolites, 51 were found in all tested conditions, with the remaining 54 in the majority of tested conditions. The 51 metabolites which were detected in all tested conditions mainly represented intermediates of the amino acid and sugar metabolism and of the tricarboxylic acid (TCA) cycle. Hence, these metabolites represent the core metabolome, which are essential for cell survival and sustainability. Among the 507 identified compounds, glutamate was found to be the most dominant compound with approximately 23% of the median peak area of all compounds. This was already described in a recent study by Frimmersdorf *et al.* (2010), where glutamate was found to be the predominant metabolite in *Pseudomonas aeruginosa* independent of the tested growth conditions, suggesting glutamate as the central storage compound for many metabolic pathways in Pseudomonads (see Figure 3.32).

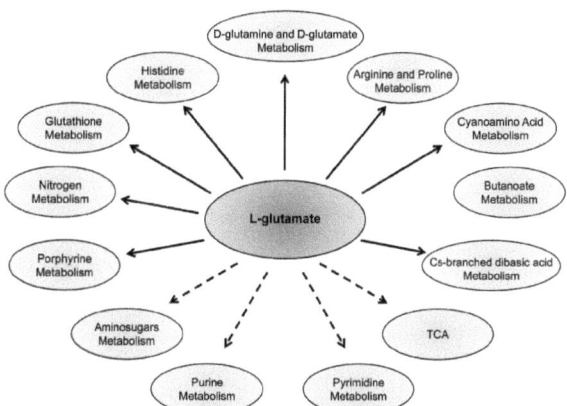

Fig. 3.32 L-glutamate as central storage compound. Metabolic routes that are fed by glutamate directly (continuous black lines) and indirectly (dashed black lines), with a few enzymatic reactions apart from the metabolic route.

Regarding the absolute metabolite concentrations of the 30°C and 10°C wild type samples, it seems as though metabolites are accumulating at 10°C, since the concentrations are generally higher at 10°C compared to 30°C. One reason for this observation might be the overall higher representation of known metabolites within the 10°C samples.

However, a few metabolites were significantly up-regulated at 10°C. Intermediates of the TCA were highly up-regulated at 10°C, mainly citrate, fumarate and malate. 2-oxo-glutarate and succinate were also identified but less up-regulated compared to the other compounds (see Figure 3.33). According to the transcriptome analysis of cold shock response of the *P. putida* KT2440 wild type, genes involved in the TCA cycle were repressed, namely *acnB* involved in the conversion of citrate to isocitrate (8.2-fold; Illumina), *ipdG*, involved in the conversion of 2-oxoglutarate to succinyl CoA (13.9-fold; Illumina), *sucC*, which converts succinyl CoA into succinate (17.4-fold; Illumina) and *sdhA*, which converts succinate into fumarate (3.6-fold; Illumina) (see Table 3.16). The combination of metabolome and transcriptome data suggests that the first part of the TCA cycle from citrate to succinate is repressed upon cold shock, leading to an accumulation of the intermediates fumarate and malate.

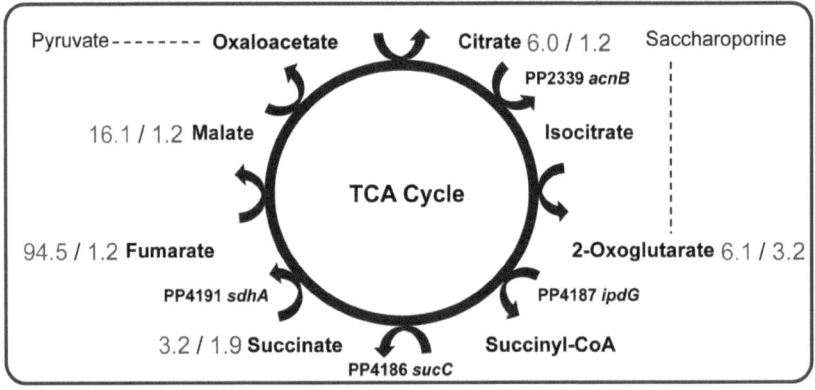

Fig. 3.33 Tricarboxylic acid cycle. Numbers indicate fold-change of respective metabolite concentrations according to the sum of total metabolite concentration in all measured samples. Blue: 10°C, red: 30°C. Genes were found to be significantly down-regulated upon cold shock in the transcriptome analysis of the *P. putida* KT2440 wild type (see Table 3.16).

Just a few metabolites dominated the absolute metabolite concentrations. The first group consisted of amino acids, which can be easily feed into the tricarboxylic acid cycle. Glutamate, as described previously in this chapter, was the most dominant compound, but the closely related amino acids alanine and aspartate were also found in high concentrations. The percentage peak area was 2.8% and 2.1%, respectively.

Furthermore, the branched-chain amino acid valine was present in high concentrations with an peak area of 7.6%, and the concentration increased following cold shock. According to the transcriptome data, valine metabolism seemed to be highly affected by cold shock. Many genes involved in the biosynthesis of valine from pyruvate, *ilvCNB* and *ilvD* were found to be repressed, as well as amino acid transporters. In contrast, genes involved in the degradation of valine were induced, such as *bkdA-2*. Thus, it is likely that valine metabolism is shifted from the biosynthesis to its degradation by repressing the reversible enzymatic reactions from valine to pyruvate, and by this facilitates its degradation to branched-chain fatty acids. In *Bacillus subtilis* it was shown that isoleucine-deficient mutants were cold sensitive, and that the biosynthesis of branched-chain fatty acids was the protective regulation upon cold shock from 37°C to 15°C (Klein *et al.*, 1999). They demonstrated thereby the important role of branched-chain fatty acids as membrane compounds and membrane fatty acids as determinants for ensuring membrane fluidity at changing temperatures. Thus, the accumulation of branched-chain amino acids is an important regulation in cold shock response by providing precursors of membrane fatty acids.

Aromatic compounds, such as benzoate and toluene, were already noted to integrate into the cell membrane and thereby alter its composition and fluidity (Sikkema *et al.*, 1995; Comes & Beelman, 2002). In contrast to aromatic compounds such as cell wall stressors, the accumulation of aromatic amino acids and derivates in high concentrations might have a protective effect on membrane physiology at low temperatures. Phenylalanine, tryptamine, as the precursor of tryptophan, and tyrosine concentrations were higher in the 10°C wild type sample. This leads to the hypothesis that aromatic amino acids might integrate into the membrane and thereby ensure its fluidity at low temperatures.

The predominant concentrations of glutamate and the remaining three amino acids suggests that these amino acids constitute the major compounds of central metabolism of Pseudomonads in a temperature independent manner by both serving as carbon and nitrogen sources, and being the central intermediates, from which all pathways can be induced. As precursors of the tricarboxylic acid cycle, they seem to accumulate at 10°C similar to the direct intermediates of the TCA cycle (see Figure 3.33).

RESULTS AND DISCUSSION

In this context another two compounds were found in high concentrations. Lactate is a strong carboxylic acid, which dissociates under physiological conditions. It can serve as carbon source, but also as a storage for the TCA cycle, since it is a direct precursor of pyruvate. One-pyrroline-3-hydroxy-5-carboxylate is an intermediate in the arginine and proline metabolism, which can be converted to pyruvate via L-erythro-4-hydroxyglutamate. In *Pseudomonas putida* the precursor of 1-pyrroline-3-hydroxy-5-carboxylate, 4-hydroxy-D-proline, has already been described to be converted into α-ketoglutarate via an oxidative reaction (Gryder & Adams, 1969, 1970). Thus, it is likely that in *P. putida* the high concentrations of 1-pyrroline-3-hydroxy-5-carboxylate constitute a storage pool for the tricarboxylic acid cycle, which is well balanced independent of environmental changes.

The second group of metabolites, which were predominant and accumulated at 10°C, were phosphorylated sugars. Fructose-, glucosamine-, glucose- and mannose-6-phosphate were present in higher concentrations at 10°C, only xylulose-5-phosphate showed no difference in concentration between the 30°C and 10°C samples. In contrast, the concentration of unphosphorylated sugars, such as maltose, cellobiose, lactose and xylulose, which all can be converted into glucose, were lower at 10°C, suggesting a preference for the synthesis of phosphorylated sugars at low temperatures. Since intermediary metabolism is overall down-regulated at 10°C, which is also reflected by a lower growth rate in general, the accumulation of intermediates of the TCA cycle and sugar metabolism might serve as a storage pool. The pentose phosphate pathway is required for glucose turnover for the production of NADPH and pentoses, which are essential compounds of nucleotides. The TCA is the final step of oxidation of carbohydrates and fatty acids. It thereby, among others, supplies NADH for metabolic processes such as oxidative phosphorylation and 2-oxoglutarate, an important precursor for many biological processes such as amino acid metabolism. Though no significant number of genes involved in the pentose phosphate pathway was found on the transcriptome level, the accumulation of phosphorylated sugars suggests that these compounds are used as energy storage, since phosphorylated sugars cannot be transferred out of the cell. Furthermore, they exhibit an osmolytic effect by binding H_2O molecules. Thus, by influencing osmolarity, they can improve transport of metabolites within the cells at low temperatures.

RESULTS AND DISCUSSION

Table 3.21 Metabolite concentrations provided as median peak* area of six technical replicates for stress sensitive Tn5 mutants.

metabolite	Number of samples found[a]	Number of samples found per condition[b]	experiment 1 cbrA::Tn5 10°C	experiment 1 cbrA::Tn5 30°C	experiment 2 cbrA::Tn5 10°C	experiment 2 cbrA::Tn5 30°C
1-monooleoylglycerol	65	1155	100.0	100.0	626.3	958.0
1-monopalmitoylglycerol	102	2366	100.0	3591.1	1000.8	1180.3
1-pyrroline-3-hydroxy-5-carboxylate	143	6666	158770.6	76979.2	45945.5	11029.3
2-keto-3-deoxygluconic acid	93	4466	41667.1	43608.9	3992.2	5425.1
2-oxoglutaric acid	93	6445	65373.7	17636.5	4032.8	1307.3
3-hydroxybutyric acid	75	2123	100.0	100.0	100.0	700.9
3-phosphoglyceric acid	94	4542	1068.3	1006.1	127.9	100.0
5'-AMP	120	6666	5752.8	8806.9	957.9	987.3
alanine	142	5666	339961.8	429738.2	70468.3	59646.6
aspartic acid	143	6666	405894.5	154930.1	80639.5	19393.5
beta-alanine	143	6666	43575.2	223065.3	7389.1	26784.9
citric acid	116	6462	9862.9	2393.3	1274.4	100.0
dodecanoic acid	125	6666	36281.7	61025.3	3947.4	7766.8
fructose 6-phosphate	136	6666	309519.3	31150.3	54603.9	6033.4
glucosamine 6-phosphate	98	4264	47372.0	100.0	9472.9	586.2
glucose	141	6666	58651.0	74557.7	46007.4	20150.0
glucose 6-phosphate	143	6666	1065421.2	122168.4	121491.4	23120.5
glutamic acid	143	6666	4515800.6	2420885.5	1097563.3	425580.8
glutamine	135	6666	104973.1	28282.5	25812.7	2809.3
glycerol	142	6666	15709.6	6429.0	1795.6	1446.6
glycerol 3-phosphate	143	6666	42525.9	30303.1	6054.7	6289.0
glycine	143	6666	66268.9	41327.1	10262.7	7659.4
homoserine	101	4466	11462.5	19655.1	2788.6	3499.0
hydroquinone	83	3366	2580.1	1713.3	923.7	928.1
inosine	119	5525	5298.2	7959.3	100.0	769.3
isoleucine	72	3516	13056.5	21691.9	100.0	8536.5
lactic acid	121	6566	258307.4	221578.5	33472.3	35998.7
lactose	88	5466	9776.9	13175.7	3928.8	3601.2
lysine	143	6666	35520.4	56205.3	7451.3	14983.2
malic acid	98	3462	357.3	757.5	176.7	100.0
maltose	98	6666	14461.0	46036.5	2551.8	8857.2
mannose 6-phosphate	117	6666	188204.8	21886.9	19371.4	3927.2
N-acetylglutamic acid	143	6666	85680.5	28845.7	15570.9	3220.3
nicotinamide	128	6565	35001.3	53224.7	6042.1	5413.7
ornithine	104	5466	25798.7	10520.6	5770.6	5055.1
oxalic acid	134	6665	34605.3	191992.3	21353.2	17392.0
palmitic acid	143	6666	124853.5	295183.4	57801.8	88380.7
phosphoethanolamine	106	3366	1517.6	2918.7	5872.4	6147.9
phosphoric acid	77	3266	526888.5	100.0	442482.1	460181.6
putrescine	143	6666	148880.2	122678.3	58472.9	46247.1
pyrophosphoric acid	86	4644	19091.7	32607.5	9607.4	5127.6
serine	135	6666	13645.4	25388.4	3503.5	5182.5
spermidine	86	3366	5571.2	1922.4	1834.7	1052.8
succinic acid	143	6666	31899.5	29254.1	7126.2	4448.1
tartaric acid	79	0066	100.0	100.0	10628.2	42728.3
threonine	143	6666	86541.9	80078.0	16930.4	11317.0
triethanolamine	77	4503	6238.5	7356.9	100.0	503.9
tyrosine	69	1404	100.0	7512.3	100.0	1240.8
uracil	101	6441	15819.5	12889.6	1692.8	100.0
valine	143	6666	1168720.3	290663.3	309099.0	50415.6
xylulose 5-phosphate	66	3433	2553.7	3352.7	543.7	454.6

Table 3.21 continued

			experiment 1		experiment 2	
metabolite	Number of samples found[a]	Number of samples found per condition[b]	cbrB::Tn5 10°C	cbrB::Tn5 30°C	cbrB::Tn5 10°C	cbrB::Tn5 30°C
1-monooleoylglycerol	65	4100	2020.5	100.0	100.0	100.0
1-monopalmitoylglycerol	102	4321	4225.6	2111.9	100.0	100.0
1-pyrroline-3-hydroxy-5-carboxylate	143	6666	53227.3	57424.0	14603.4	9717.5
2-keto-3-deoxygluconic acid	93	2412	100.0	13127.1	100.0	100.0
2-oxoglutaric acid	93	6212	7758.0	100.0	100.0	100.0
3-hydroxybutyric acid	75	0010	100.0	100.0	100.0	100.0
3-phosphoglyceric acid	94	2543	100.0	517.2	95.6	68.8
5'-AMP	120	6666	1246.7	2419.8	492.1	300.0
alanine	142	6666	62666.9	201974.0	28929.4	28573.1
aspartic acid	143	6666	168164.9	82102.8	21953.8	6559.8
beta-alanine	143	6666	11671.3	86494.4	7643.0	26323.3
citric acid	116	5352	1596.4	932.9	771.8	100.0
dodecanoic acid	125	6666	6614.9	25984.0	6121.5	6816.2
fructose 6-phosphate	136	6564	17540.3	3517.8	3799.1	653.6
glucosamine 6-phosphate	98	3130	1236.3	100.0	73.2	100.0
glucose	141	6666	9934.3	18961.1	1401.3	1513.2
glucose 6-phosphate	143	6666	82682.7	23230.6	8325.8	2745.7
glutamic acid	143	6666	2993071.5	2195864.5	660200.1	420764.2
glutamine	135	6665	21875.1	9030.4	4291.4	776.9
glycerol	142	6666	5146.5	4532.4	1145.7	618.3
glycerol 3-phosphate	143	6666	17827.1	17573.2	2826.6	2713.1
glycine	143	6666	18994.0	18830.0	9454.7	4292.6
homoserine	101	3433	3256.6	11468.6	830.2	416.2
hydroquinone	83	1155	100.0	100.0	786.9	811.6
inosine	119	2656	100.0	3299.5	379.6	522.8
isoleucine	72	3514	3006.6	21681.2	100.0	2922.8
lactic acid	121	5365	45609.3	44623.2	22111.5	9786.7
lactose	88	4566	11411.0	43702.9	5629.1	5872.3
lysine	143	6666	5317.6	37609.1	5158.5	4192.4
malic acid	98	2300	100.0	75.8	100.0	100.0
maltose	98	6666	26734.8	144122.7	223790.3	206143.7
mannose 6-phosphate	117	6362	11944.0	1469.1	1595.0	100.0
N-acetylglutamic acid	143	6666	13215.5	5776.6	5040.2	1347.8
nicotinamide	128	4612	5622.1	20182.5	100.0	100.0
ornithine	104	5264	8463.5	100.0	6076.4	1224.0
oxalic acid	134	6566	25327.6	34239.7	18497.0	17568.7
palmitic acid	143	6666	132414.6	183251.0	46725.5	48726.1
phosphoethanolamine	106	5565	11040.7	10706.6	3424.2	2728.5
phosphoric acid	77	4451	1241431.2	1575350.0	294365.6	100.0
putrescine	143	6666	27666.8	139925.0	65969.6	60500.1
pyrophosphoric acid	86	6553	36142.6	14923.6	4391.0	747.0
serine	135	6666	4892.9	16108.1	2368.0	1175.0
spermidine	86	2320	100.0	1962.6	100.0	100.0
succinic acid	143	6666	9583.9	10233.1	3186.7	1042.4
tartaric acid	79	0461	100.0	6593.1	3218.0	100.0
threonine	143	6666	24999.5	48275.3	6439.1	5990.6
triethanolamine	77	5566	2641.3	5586.2	1185.6	1180.2
tyrosine	69	0300	100.0	1937.9	100.0	100.0
uracil	101	6530	16628.8	7283.8	559.0	100.0
valine	143	6666	43522.5	118512.5	13726.3	16903.1
xylulose 5-phosphate	66	1300	100.0	1053.3	100.0	100.0

Table 3.21 continued

metabolite	Number of samples found[a]	Number of samples found per condition[b]	experiment 1		experiment 2	
			pcnB::Tn5 10°C	pcnB::Tn5 30°C	pcnB::Tn5 10°C	pcnB::Tn5 30°C
1-monooleoylglycerol	65	5300	2888.0	629.4	100.0	100.0
1-monopalmitoylglycerol	102	5610	4971.7	3717.6	100.0	100.0
1-pyrroline-3-hydroxy-5-carboxylate	143	6666	231057.0	62226.0	36056.0	11125.6
2-keto-3-deoxygluconic acid	93	0043	100.0	100.0	7850.8	2483.1
2-oxoglutaric acid	93	0054	100.0	100.0	5819.5	1379.0
3-hydroxybutyric acid	75	6063	17798.3	100.0	12866.8	2497.6
3-phosphoglyceric acid	94	6541	508.5	228.3	247.8	100.0
5'-AMP	120	5666	614.3	1136.6	870.5	993.5
alanine	142	6666	472767.2	170589.4	157652.4	48010.0
aspartic acid	143	6666	330588.5	117265.3	105819.0	12338.5
beta-alanine	143	6666	26225.0	35116.5	13879.1	21861.3
citric acid	116	6242	2057.6	100.0	674.7	100.0
dodecanoic acid	125	4366	3493.1	996.3	4381.1	11345.3
fructose 6-phosphate	136	6465	6937.8	371.2	120017.6	3988.3
glucosamine 6-phosphate	98	6662	94870.4	2813.2	11007.1	100.0
glucose	141	6666	155351.1	95034.9	5568.1	3595.9
glucose 6-phosphate	143	6666	263220.6	10225.3	261843.3	11176.6
glutamic acid	143	6666	2292404.1	800010.9	1039830.3	255566.6
glutamine	135	6665	93622.1	17721.9	54262.9	3695.7
glycerol	142	6666	4814.3	2377.9	3026.8	1963.6
glycerol 3-phosphate	143	6666	30909.0	20799.8	15463.9	5737.5
glycine	143	6666	39618.6	138727.5	17205.3	11404.6
homoserine	101	6665	11934.0	9465.6	4627.8	2194.7
hydroquinone	83	6445	3718.2	2893.1	1071.5	1124.9
inosine	119	6656	10395.8	7614.7	1599.4	3143.7
isoleucine	72	5314	31662.2	3342.5	100.0	5361.8
lactic acid	121	6663	136748.2	57600.3	36313.8	1395.2
lactose	88	1020	100.0	100.0	100.0	100.0
lysine	143	6666	65927.6	48778.9	33022.7	11735.6
malic acid	98	6552	588.9	322.5	196.9	100.0
maltose	98	0000	100.0	100.0	100.0	100.0
mannose 6-phosphate	117	0064	100.0	100.0	45970.3	1475.2
N-acetylglutamic acid	143	6666	52197.0	32210.4	19821.5	6838.3
nicotinamide	128	6655	24243.5	21935.0	10402.4	8066.1
ornithine	104	6643	7875.9	5294.2	2268.0	741.8
oxalic acid	134	5465	5971.6	2839.7	32569.1	32314.4
palmitic acid	143	6666	205752.7	231877.7	55021.0	41246.9
phosphoethanolamine	106	3462	477.7	1276.5	4535.3	100.0
phosphoric acid	77	0465	100.0	882151.1	677167.9	523745.8
putrescine	143	6666	158075.0	129959.3	58072.2	29413.8
pyrophosphoric acid	86	1224	100.0	100.0	100.0	2182.7
serine	135	6666	12424.7	25498.2	4421.8	5223.8
spermidine	86	6643	4926.8	2319.9	1599.3	598.9
succinic acid	143	6666	31678.0	13936.7	9009.6	4440.5
tartaric acid	79	5636	15233.6	17757.4	1887.8	7433.1
threonine	143	6666	93703.5	28584.6	38369.5	10705.2
triethanolamine	77	0023	100.0	100.0	100.0	470.7
tyrosine	69	6511	11993.1	4634.1	100.0	100.0
uracil	101	6644	12005.5	16416.0	1875.7	2812.2
valine	143	6666	1223882.4	193694.7	848688.4	62411.4
xylulose 5-phosphate	66	0031	100.0	100.0	354.4	100.0

Table 3.21 continued

			experiment 1		experiment 2	
metabolite	Number of samples found[a]	Number of samples found per condition[b]	vacB::Tn5 10°C	vacB::Tn5 30°C	vacB::Tn5 10°C	vacB::Tn5 30°C
1-monooleoylglycerol	65	5255	2713.1	100.0	1539.7	3285.9
1-monopalmitoylglycerol	102	6665	5818.0	3191.4	3118.1	5493.3
1-pyrroline-3-hydroxy-5-carboxylate	143	6666	102409.6	60110.1	107179.5	22910.8
2-keto-3-deoxygluconic acid	93	6666	30440.1	24139.3	30668.5	26247.7
2-oxoglutaric acid	93	4551	5681.9	9830.0	11872.5	100.0
3-hydroxybutyric acid	75	6465	22480.1	6988.8	64390.7	6288.2
3-phosphoglyceric acid	94	5430	281.0	259.3	161.3	100.0
5'-AMP	120	3400	73.4	682.3	100.0	100.0
alanine	142	6666	392986.5	146656.9	348008.1	108973.7
aspartic acid	143	6666	177219.4	173452.2	130708.9	35882.6
beta-alanine	143	6666	52618.1	63693.1	42354.0	65228.0
citric acid	116	6665	2368.2	1940.8	3901.3	1512.7
dodecanoic acid	125	6666	30433.2	23795.4	34395.3	25501.9
fructose 6-phosphate	136	6666	369388.0	14952.5	188643.3	12142.8
glucosamine 6-phosphate	98	6260	24970.2	100.0	6149.2	100.0
glucose	141	6666	38431.6	18463.4	55982.1	25772.4
glucose 6-phosphate	143	6666	718482.4	51646.2	504275.8	31401.2
glutamic acid	143	6666	2360114.7	1142344.4	2182758.0	580555.8
glutamine	135	6664	133771.8	38782.1	71110.0	2063.1
glycerol	142	6566	5154.4	4777.6	6107.8	4371.6
glycerol 3-phosphate	143	6666	31878.4	15626.8	23411.4	10732.9
glycine	143	6666	56540.0	30071.2	44583.1	33927.8
homoserine	101	6666	17642.4	13897.5	16302.6	10622.3
hydroquinone	83	4433	2080.4	1810.8	1057.0	1051.4
inosine	119	5606	5466.5	5582.5	100.0	1556.0
isoleucine	72	4365	27769.2	2732.0	60531.5	7205.9
lactic acid	121	5254	92135.1	100.0	56373.9	14299.4
lactose	88	1065	100.0	100.0	8633.3	4798.6
lysine	143	6666	92382.0	38356.7	110735.6	38273.6
malic acid	98	6652	266.1	320.1	395.8	100.0
maltose	98	6666	3094.5	6161.5	7660.0	3704.8
mannose 6-phosphate	117	6665	155250.6	9753.8	108943.3	5680.4
N-acetylglutamic acid	143	6666	33358.2	74562.2	34953.8	15540.4
nicotinamide	128	6666	27135.0	23711.0	36662.0	27270.0
ornithine	104	6466	7076.0	2714.7	12606.8	10919.3
oxalic acid	134	6666	77514.0	90318.3	209091.7	151360.8
palmitic acid	143	6666	284206.3	217175.7	143222.9	213536.8
phosphoethanolamine	106	6644	26360.7	4389.1	3855.6	5700.7
phosphoric acid	77	2166	100.0	100.0	1165193.5	918007.0
putrescine	143	6666	201973.6	158940.9	201431.4	179099.6
pyrophosphoric acid	86	0235	100.0	100.0	11682.3	17671.0
serine	135	6666	8355.3	19067.6	21176.7	16860.0
spermidine	86	2616	100.0	5073.5	100.0	7875.4
succinic acid	143	6666	28360.9	17357.7	26903.7	17395.1
tartaric acid	79	3466	11805.5	20427.4	10367.3	14693.0
threonine	143	6666	82605.6	39903.4	75326.4	26829.7
triethanolamine	77	5665	4058.9	5657.6	4690.7	3498.5
tyrosine	69	6362	16645.9	867.1	21102.6	100.0
uracil	101	6612	8035.0	22269.4	100.0	100.0
valine	143	6666	961615.6	340027.8	1516509.3	201608.7
xylulose 5-phosphate	66	1601	100.0	2506.3	100.0	100.0

Table 3.21 continued

			experiment 1		experiment 2	
metabolite	Number of samples found[a]	Number of samples found per condition[b]	bipA::Tn5 10°C	bipA::Tn5 30°C	bipA::Tn5 10°C	bipA::Tn5 30°C
1-monooleoylglycerol	65	5210	4915.4	100.0	100.0	100.0
1-monopalmitoylglycerol	102	5652	9109.9	5925.9	1398.7	100.0
1-pyrroline-3-hydroxy-5-carboxylate	143	5666	152416.3	59333.4	47756.1	15511.7
2-keto-3-deoxygluconic acid	93	4212	16610.9	100.0	100.0	100.0
2-oxoglutaric acid	93	3265	24037.5	100.0	9144.0	2928.9
3-hydroxybutyric acid	75	0060	100.0	100.0	9351.0	100.0
3-phosphoglyceric acid	94	5134	470.6	100.0	163.1	163.6
5'-AMP	120	5266	2686.1	100.0	1317.6	1221.7
alanine	142	5666	477667.9	127205.2	189026.4	48476.0
aspartic acid	143	5666	472217.2	111103.7	189222.7	33040.6
beta-alanine	143	5666	34758.1	58448.9	17902.1	27398.5
citric acid	116	5665	2578.7	2011.7	3240.0	810.3
dodecanoic acid	125	5666	17188.3	10919.1	7384.0	9039.3
fructose 6-phosphate	136	5566	443321.3	9809.9	97756.6	7732.4
glucosamine 6-phosphate	98	4463	25100.8	2354.8	12073.9	71.7
glucose	141	5565	30975.5	9043.2	5071.3	1969.3
glucose 6-phosphate	143	5666	730925.0	40437.3	208405.0	22026.1
glutamic acid	143	5666	1525624.4	1004269.3	1643895.5	455934.1
glutamine	135	5666	128946.0	6397.7	57745.7	5235.9
glycerol	142	5666	29305.4	63888.8	2317.2	1898.0
glycerol 3-phosphate	143	5666	25841.4	10109.0	13936.4	5342.3
glycine	143	5666	84054.2	18795.2	27786.7	16894.4
homoserine	101	5655	14146.8	10589.5	3771.2	5198.8
hydroquinone	83	5645	2872.0	3155.9	1493.0	1255.3
inosine	119	4656	7338.5	9802.6	2109.1	3113.5
isoleucine	72	4601	34444.9	10548.4	100.0	100.0
lactic acid	121	5364	117741.9	8809.6	46253.5	7254.3
lactose	88	0044	100.0	100.0	2101.3	1288.6
lysine	143	5666	81405.4	41529.6	27069.9	15580.8
malic acid	98	5554	498.8	252.3	224.3	75.6
maltose	98	1100	100.0	100.0	100.0	100.0
mannose 6-phosphate	117	4366	110677.9	6202.8	34088.5	4175.0
N-acetylglutamic acid	143	5666	71253.7	31710.4	26640.2	9498.0
nicotinamide	128	5666	28842.9	29887.5	8888.3	8673.7
ornithine	104	5031	9892.9	100.0	1248.9	100.0
oxalic acid	134	4466	65888.6	51230.6	54088.8	41317.4
palmitic acid	143	5666	343586.3	329273.3	101403.0	82590.5
phosphoethanolamine	106	5343	22527.3	777.1	7302.6	2204.9
phosphoric acid	77	0365	100.0	340831.4	821365.7	575142.8
putrescine	143	5666	147938.6	114906.4	44795.0	53281.7
pyrophosphoric acid	86	4034	5376.8	100.0	2028.2	14020.8
serine	135	5666	21844.1	23522.4	11525.8	9440.9
spermidine	86	3223	1488.6	100.0	100.0	422.4
succinic acid	143	5666	36762.0	13139.6	11132.1	4624.4
tartaric acid	79	4364	14290.7	2063.3	20752.7	4452.4
threonine	143	5666	105858.2	48863.0	50069.3	18382.5
triethanolamine	77	4353	4531.7	2133.2	1975.1	646.6
tyrosine	69	5562	11109.9	6419.8	6699.0	100.0
uracil	101	5665	21425.9	20761.7	4083.9	2825.4
valine	143	5666	982800.5	188049.4	841311.5	76348.3
xylulose 5-phosphate	66	3263	6392.7	100.0	4933.3	570.4

[a] Total number of samples, in which the respective metabolite was detected (in total 143 samples were analyzed).

[b] Number of samples per strain and condition, in which the metabolite was detected (6 technical replicates per condition, in order 10°C, 30°C of experiment 1; 10°C, 30°C of experiment 2). For experiment 1 only 5 samples of the 10°C condition were analyzed.

* The median peak area was calculated from the raw data provided by the GC-MS analysis. 100.0 means the metabolite was not detected.

RESULTS AND DISCUSSION

The above described metabolome profile was based on the metabolite concentrations of the *P. putida* KT2440 wild type upon cold shock, and differences in the 30°C and 10°C samples were considered to represent the cold shock response on the metabolite level. When comparing the metabolite concentrations of the respective mutants, the results show that the mutants, mainly reflected by the metabolome profile of the *cbrB*::Tn5 mutant, are affected in the central metabolic routes, which were differentially regulated upon cold shock. This was reflected by the predominance of blue in the *cbrB*::Tn5 mutant profile (see Figure 3.31), which indicated that metabolites of the core metabolome are generally reduced in this mutant compared to the wild type.

For the comparison of the metabolome profiles of wild type and Tn5 mutants, a principle component analysis was applied. Principle component analysis is a mathematical tool to transform a data set into a new coordinate system, whereby the greatest variance is displayed on the first coordinate. This can be used for the visualization of variance in a complex data set. Here, a simplified PCA was applied including the logarithmized median metabolite concentrations (combined from experiment 1 and 2) per condition. For the PCA the dChip and Tigr MEV (Multiple Experimental Viewer) software (http://biosun1.harvard.edu/complab/dchip/; http://www.tm4.org/mev/) were used, including the complete list of all metabolites. The results showed that 51% of the total variance was attributed to the first two components. Since many unknown metabolites contributed to the majority of either component 1 or 2, the analysis was repeated with a reduced list including only known metabolites. Still, nearly 50% of the total variance could be attributed to component 1 and 2. A list of the first 20 metabolites that contributed to 54% to component 1 and 56% to component 2 is provided in Table 3.22. The PCA plot (Figure 3.34) clearly shows that the metabolite profile of the *P. putida* KT2440 wild type and the respective mutants are distinct from each other, as indicated by the distance calculated for principle component 1.

The distance between wild type and mutants (component 1) is greater than the distance between the two conditions (30°C and 10°C) (component 2). Nevertheless, the distance between the 30°C and 10°C samples is significant, indicating a temperature dependent metabolite profile of all strains, apart from the *cbrB*::Tn5 mutant. Only the 30°C and 10°C samples of the *cbrB*::Tn5 mutant clustered. This suggests that the metabolite profile of the two conditions is similar, and that the *cbrB*::Tn5 mutant has an impaired cold shock response.

RESULTS AND DISCUSSION

Table 3.22 Principle component analysis. PC 1 accounts for 27.9% and PC 2 for 21.8% of the total observed variance. The list displays the first 20 metabolites, which account for 54% of component 1 and 56% of component 2.

	component 1	27.90%	component 2	21.80%
1	4.31%	proline	6.11%	pyruvic acid
2	3.92%	ribulose 5-phosphate	5.49%	maltose
3	3.64%	5-oxoproline	4.50%	ribose
4	3.45%	phenylalanine	3.56%	oxamic acid
5	2.90%	tartaric acid	3.07%	lactose
6	2.84%	4-aminobutanoic acid	3.00%	1,3-diphosphoglyceric acid
7	2.74%	isoleucine	2.82%	phenylalanine
8	2.73%	homoserine	2.80%	sucrose
9	2.69%	triethanolamine	2.54%	isoleucine
10	2.62%	mannitol	2.42%	proline
11	2.55%	fumaric acid	2.21%	6-phosphogluconic acid
12	2.51%	cytosine	2.19%	tartaric acid
13	2.46%	xylulose	2.10%	homoserine
14	2.33%	fructose	2.09%	tryptophan
15	2.23%	carbonic acid	2.04%	2-keto-3-deoxygluconic acid
16	2.19%	2-hydroxyglutaric acid	2.02%	ethanolamine
17	2.13%	TMP	1.96%	5-oxoproline
18	2.03%	N-acetylglucosamine	1.94%	pyrophosphoric acid
19	1.88%	oxamic acid	1.87%	4-aminobutanoic acid
20	1.84%	tryptamine	1.69%	benzoic acid
	53.98%		56.42%	

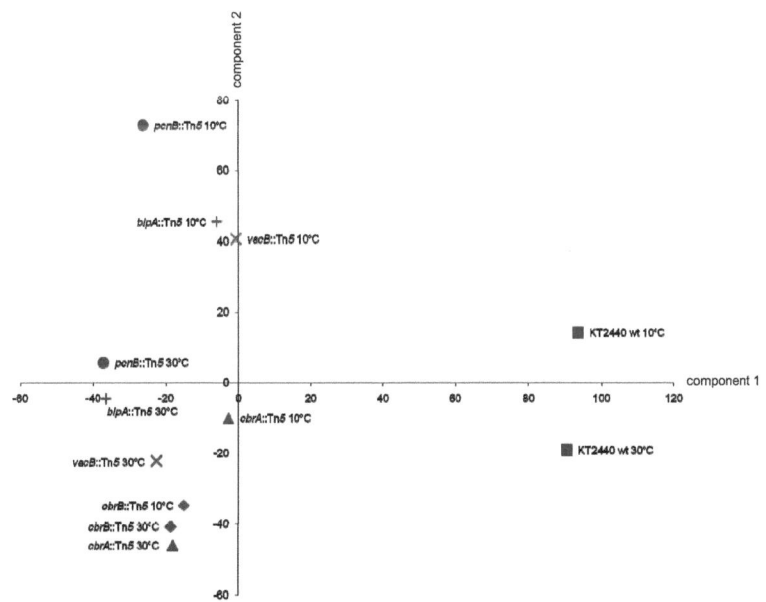

Fig. 3.34 PCA (principle component analysis) plot of logarithmically transformed median metabolite concentrations from *P. putida* KT2440 wild type and the five cold sensitive Tn5 mutants at two different temperatures.

The first 20 metabolites, which together contributed about 54% of component 1, consisted mainly of branched-chain and aromatic amino acids and derivates and of intermediates of the pentose phosphate pathway and tricarboxylic acid cycle. According to the wild type metabolome profile, the accumulation of intermediates of the pentose phosphate pathway and TCA cycle at 10°C was already described previously to may serve as storage pool, for instance for the production of NADPH. The distance between wild type and Tn5 mutants as calculated by the PCA for component 1 thus reflects a temperature independent impairment of central metabolism of the mutants, which results in the stress sensitive phenotype. Interestingly, according to component 1, all components representative for the Tn5 mutants are located close to each other. This indicates that the Tn5 mutants, though affected in independent genes of diverse functionality, exhibit a similar metabolome profile, which is distinct from the wild type.

The metabolome profile supports the previous findings that cbrB-deficient mutants are impaired in utilizing certain amino acids. Here, all five mutants were impaired in catabolizing proline as reflected by its accumulation. In addition, further metabolites of the arginine and proline metabolism were affected in the stress sensitive mutants. This was most visible in the metabolome profile of the cbrB::Tn5 mutant, where glutamine, N-acetyl glutamate and ornithine were hardly detected in the 30°C samples, and were only slightly increased in the 10°C samples. Furthermore, 4-aminobutanoate was only detected in high concentrations in the wild type, but not in the mutants (only in minor concentrations in experiment 1 in the 10°C samples of the cbrB::Tn5 mutant). Since proline is apparently one of the preferentially utilized amino acids in P. putida, the defect of the mutants (mainly reflected by the cbrB::Tn5 profile) in proficiently using the arginine and proline metabolic pathway, can explain the stress sensitive phenotype of these mutants.

In the literature it is well described that cbrAB-deficient mutants are unable to utilize several amino acids, such as histidine, proline, leucine and isoleucine and polyamines (Nishijyo et al., 2001; Zhang & Rainey, 2008). Indeed, the metabolome profile of the stress sensitive mutants showed an accumulation of the amino acids proline and isoleucine. Although proline was described to be a more preferred amino acid than isoleucine in P. aeruginosa (Frimmersdorf et al., 2010), it seems as these amino acids are completely metabolized by the P. putida wild type, whereas the stress sensitive mutants are incapable of utilizing isoleucine and proline leading to their accumulation in the mutant background. Furthermore, the two compounds homoserine and hydroquinone, which are intermediates in cysteine and tyrosine metabolism, respectively, were only detected in the mutants but not in the wild type, supporting the results of altered amino acid utilization in the mutants. Interestingly, the compound 5-oxoproline was only found in the wild type sample, but not in the mutants (besides the *vacB*::Tn*5* 30°C sample). 5-oxoproline is an intermediate in the degradation pathway of glutathione to glutamate. Gluthathione is transferred outside of the cell membrane, where the γ-glutamyl group is transferred from glutathione to an external amino acid. The γ-glutamyl amino acid is then transferred into the cell, where it is converted via 5-oxoproline to glutamate. In the latter step, ATP is hydrolyzed. Since the mutants seemed to be affected in glycolysis and TCA, where ATP is generated, it is likely that the mutants avoid the synthesis of glutamate from glutathione to reduce ATP-dependent reactions.

Regarding the TCA, where fumarate, malate and citrate concentrations were strongly induced at 10°C, the *cbrB*::Tn*5* mutant profile exhibited no significant increase in metabolite concentrations upon cold shock. Indeed, the concentrations of intermediates of the TCA were lower compared to the wild type and the other mutants. Fumarate, which was highly induced at 10°C in the wild type, was hardly found in any of the mutant samples, apart from the 10°C sample of the *vacB*::Tn*5* mutant. Besides fumarate, tartaric acid was found to be among the first 20 metabolites, which mainly attributed to component 1. Tartrate is a precursor of oxaloacetate, an intermediate of the TCA cycle. This compound was not found in the wild type, but in the Tn*5* mutants, in contrast to fumarate as well as pyruvate, which was found in higher concentrations in the wild type. This suggests that the mutants may compensate a defect in TCA cycle by using alternative pathways, which also lead to the production and conversion of TCA intermediates.

Many metabolites, which contribute to component 2 and thus reflect the distinguished metabolome profile following cold shock, were also found in component 1. Hence, the impairment in central metabolism of the Tn*5* mutants according to component 1 leads to the cold sensitive phenotype of the mutants, since the majority of metabolites contributing to component 1 are also involved in cold shock response.

Pyruvate was the major metabolite which contributed about 6% to component 2. The PCA clearly showed that metabolites of component 2 reflect the different metabolite profiles following cold shock. The distances between the 30°C and 10°C samples were even more pronounced by the Tn*5* mutants, apart from the *cbrB*::Tn*5* mutant. This demonstrates that the impairment of central metabolism can be partially compensated in the remaining four mutants, which leads to a similar PCA pattern as the wild type according to the distribution along component 2.

The concentration of maltose was significantly higher in the *cbrB*::Tn*5* mutant compared to all other strains, and not found in the *pcnB*::Tn*5* and *bipA*::Tn*5* mutants. This finding suggests, since *cbrB*-deficient mutants are known to exhibit growth defects when xylose, mannose, ribose and arabitol, among others, are used as sole carbon source (Zhang & Rainey, 2008), that the *cbrB*::Tn5 mutant attempts to compensate the defect in sugar metabolism by accumulating maltose as a central compound for sugar metabolism.

In contrast to the accumulation of metabolites at low temperatures, a few compounds were found in lower concentrations upon cold shock. Concentrations of central compounds of the purine and pyrimidine metabolism, inosine, adenosine 5'-monophosphate and cytosine, respectively, and nicotinamide of the nicotinate and nicotinamide metabolism were decreased. This reflects the findings of the transcriptome analysis, as key genes for nucleotide biosynthesis were down-regulated following cold shock. Furthermore, ethanolamine and hydroxylamine, which are central compounds in glycerophospholipid and nitrogen metabolism, respectively, were found in lower concentrations at 10°C. The reduction of hydroxylamine upon cold shock seems to be an important mechanism, since this was found consistently in all strains, except for the *vacB*::Tn5 mutant.

Furthermore, the phosphoglycerolipid metabolism seemed to be strongly affected in the stress sensitive mutants compared to the wild type. Glycerol 3-phosphate was present in higher concentrations at 10°C in the wild type. Though, the concentrations of the mutants was generally lower compared to the wild type, only the *cbrB*::Tn5 mutant exhibited no differences in concentration between the 30°C and 10°C samples. Phospho- and triethanolamine were less apparent in the wild type, but present in higher concentrations in the mutants, only the *pcnB*::Tn5 mutant showed comparable concentrations to the wild type. Ethanolamine in contrast, was present in high concentrations in the 30°C of the wild type, but decreased upon cold shock, whereas this metabolite was only found in the *vacB*::Tn5 mutant in comparable amounts. Phosphoglycerolipids are important compounds of the cell membrane. Since a decrease in temperature has a major impact on the cell membrane by affecting membrane fluidity (Phadtare, 2004), the effects on phosphoglycerolipid metabolism in the mutants' background might explain the cold sensitivity.

Metabolome studies, as the third component in the functional genomics field enable the comprehensive analysis of the metabolite profile to changing growth conditions. In this study, the metabolome profile in response to cold shock of the *P. putida* KT2440 wild type in comparison to the stress sensitive mutants was examined. It provided thereby a complementary view to the transcriptome data. Indeed, the down-regulation of central metabolic routes upon cold shock, as indicated by the transcriptome results, were supported by the identification of key metabolites in the respective pathways. The down-regulation of the TCA cycle was reflected by genes involved in the different conversion steps from citrate to fumarate on the one hand, and the accumulation of fumarate, malate and citrate on the other hand. Though no genes were found to be differentially regulated upon cold shock in the mutants' background, the metabolome profile clearly showed that the mutants are affected in arginine and proline metabolism, mainly reflected by the *cbrB*::Tn5 mutant. Since the stress sensitive mutants showed a different pattern in the metabolite profile compared to the wild type of these central metabolic routes, this might be one reason for the stress sensitive phenotype. The cold sensitive phenotype was furthermore supported by the identification of metabolites of the phosphoglycerolipid metabolism. As phosphoglycerolipids are important compounds of the cell membrane, and membrane fluidity is impaired at low temperatures, an intact regulation is required for a proficient cold shock response.

The synthesis of branched-chain and aromatic amino acids was already postulated to be an important mechanism in cold shock response due to the transcriptome data of the wild type and stress sensitive mutants (see chapter 3.7 and 3.8). Indeed, the metabolome data here supported the hypothesis that these amino acid groups are important in cold shock response, as the concentrations were increasing at 10°C. In combination with the transcriptome data, the valine metabolism seemed to be shifted from its synthesis to its degradation to ensure valine as precursor for the synthesis of branched-chain fatty acids. Klein *et al.* (1999) already showed that branched-chain amino acids are required for maintenance of membrane fluidity at low temperatures. The transcriptome data of the stress sensitive mutants revealed that the *ped* gene cluster, which was described in *P. putida* U (Arias *et al.*, 2008) to be required for the conversion of 2-phenylethanol, was consistently repressed in all five mutants independent of the temperature. The transcriptome data of the wild type showed that the regulation of branched-chain amino acids was strongly affected upon cold shock, and the metabolome data supported this finding, as the concentrations of aromatic and branched-chain amino acids were increased at low temperatures. The mutants showed a different metabolite profile of these compounds compared to the wild type, which supports the hypothesis that the *ped* cluster is involved in the synthesis of branched-chain and aromatic amino acids, resulting in the cold sensitive phenotype.

RESULTS AND DISCUSSION

The highly ranked regulatory function of *cbrB* in Pseudomonads was supported by the metabolite profile of the mutant. The transcriptome data already revealed that many genes that are required for cold shock response are *cbrB*-dependent, and thus leading to cold sensitivity in a *cbrB*-negative background. Furthermore, the PCA revealed that the metabolite profiles of the *cbrB*::Tn5 mutant were nearly unaffected upon cold shock, which resulted in clustering of the two samples. Though, all mutants showed a partially different pattern of certain metabolites, the difference between the wild type and the *cbrB*::Tn5 mutant was strongest. This suggests that, though all five mutants are similarly impaired in central metabolism according to the PCA, a mutation in *cbrA*, *pcnB* and *vacB* can be partially compensated, if *cbrB* is intact.

4 PERSPECTIVES

In the present study the response of the industrially important strain *Pseudomonas putida* KT2440 to selected stress conditions, namely high concentrations of sodium benzoate and a sudden decrease in temperature were examined. In the analysis of cold shock five stress sensitive mutants, which were affected in the genes *cbrA*, *cbrB*, *pcnB*, *vacB* and *bipA*, respectively, were included in the experiments to reveal key mechanisms in stress response. For this, a functional genomics approach was used, integrating transcriptome, proteome and metabolome data for a comprehensive analysis of the stress profile of the *P. putida* wild type and selected mutants.

The benzoate stress experiments were carried out in collaboration with other members of the "PSYSMO" consortium, and are still in progress at the time of writing.
The preliminary gene expression analysis of the benzoate pulse implementation experiment already revealed a two-phase response to a sudden increase of benzoate. In the first phase regulatory networks were induced, such as the benzoate degradative ß-ketoadipate pathway and aromatic specific transporter systems, which help to reduce the benzoate concentration within the cells. In the second phase, genes associated with cell wall stress were induced, reflecting a kind of repair mechanism. The systems biology approach utilized here to feed mathematical models of regulatory networks with the complementary results from the different functional genomics fields might reveal new regulatory networks in the response to high concentrations of aromatic compounds. Since the bioproduction of fine chemicals, such as p-hydroxystyrene and toluene from glucose is of industrial interest (Wierckx *et al.*, 2005; Verhoef *et al.*, 2009), the identification of novel regulatory pathways can assist in genetic manipulation and improvement of the strain for biotechnological applications.
In another study the adaptation processes of *P. putida* KT2440 to changing carbon sources, from benzoate to glucose to fructose and back to benzoate are under investigation. Though the results from metabolic flux and transcriptome analysis are still pending at time of writing, these experiments promise to reveal the metabolic versatility of *P. putida* in more detail.

With the introduction of deep sequencing technologies, genomic research has changed drastically, since they allow addressing more global biological questions in greater depth. In this study the Illumina deep sequencing technology was applied to analyse the transcriptome profile of the *P. putida* KT2440 wild type upon cold shock. Besides the comparison with results from two DNA microarray platforms (Progenika and Affymetrix) and the integration of the results derived from the three transcriptome platforms to determine the key players in cold shock response, the transcriptome data from the Illumina cDNA sequencing revealed a high number of yet undescribed transcripts. In total, 143 hypothetical proteins in 105 intergenic regions, and another 66 unclassified transcripts, which are either protein or sRNA coding, were identified. Furthermore, 22 sRNAs could be verified in the *P. putida* genome due to sequence similarity to already described sRNA genes in other *Pseudomonas* species. These results demonstrated the great potential of this approach to reveal yet undiscovered coding regions. With the identification of the complete coding regions including those for sRNAs, the understanding of regulatory networks of *P. putida* will be significantly improved. These networks facilitate the great metabolic diversity and stress-tolerance to aromatic compounds, among others, of this strain and thus may help to improve the strain for industrial applications. Considering this more globally, the elucidation of the entire regulatory networks of bacteria and understanding its interconnections, enables the identification of new targets, which might be susceptible for novel antibiotics and drugs in the medical field.

The transcriptome analysis of the five cold sensitive Tn5 mutants, affected in *cbrA*, *cbrB*, *pcnB*, *vacB* and *bipA*, supported the results of previous studies, that hypothesized CbrB to be a highly ranked regulator in the regulatory cascades of Pseudomonads (Sonnleitner *et al.*, 2009; Zhang *et al.*, 2010). The transcriptome and metabolome profiles upon cold shock of the stress sensitive mutants showed the strongest influence on gene expression and metabolite concentration in the *cbrB*-negative background. This suggested that defects in the four remaining genes can be somehow compensated by other regulatory mechanisms. Indeed, the transcriptome data of this study showed a *cbrB*-dependent expression of *vacB*, and *pcnB*-expression was previously reported to be *cbrB*-dependent (Zhang *et al.*, 2010). Since CbrB is well characterized to be required for a healthy C::N balance, utilization of certain amino acids (Nishijyo *et al.*, 2001; Zhang & Rainey, 2008), and by regulation of the *crcZ* sRNA gene to be involved in catabolite repression control (Sonnleitner *et al.*, 2009), it will be interesting to identify more target genes of CbrB to elucidate its entire regulatory potential. In a future collaboration with another partner from the "PSYSMO" consortium CbrB will be used for protein-protein interaction experiments to identify potential reaction partners of CbrB.

5 CONCLUSION

The capability of *Pseudomonas putida* KT2440 as a biosafety strain to survive and function in the environment makes it a promising candidate as a model organism for genetic modifications to alter and improve regulatory networks that enable the application in biotechnological processes. Systems biology is an approach to understand the cell as a biological system as a whole and to elucidate all regulatory networks and interactions that are involved in response to certain tested conditions. To achieve this, a combination of different experimental setups, e.g. the combination of different "omics", and an integration of these experimental driven data into mathematical models is essential to improve computationally driven predictions that can then again help with the design of future experiments. From the experimental side, the screening of transposon mutant libraries has been shown to be a successful approach to identify key genes that are required for growth under certain tested conditions, e.g. nutrient limitation, temperature or high concentrations of xenobiotics.

In this study, Tn5 transposon mutants, previously described by Reva *et al.* (2006) were screened for correct plasposon insertion and screened for stress sensitive phenotypes. Out of 32 tested Tn5 mutants, five were selected for a combined transcriptome and metabolome analysis of cold shock response in comparison to the wild type response. The key players of cold shock response of the *P. putida* wild type were determined by a comprehensive analysis of three transcriptome platforms, Illumina cDNA sequencing and Affymetrix and Progenika DNA microarrays. Proteome and metabolome data complemented the transcriptional profile upon cold shock. Furthermore, the cDNA sequencing results were used for expression analysis of intergenic regions.

The first part of this study dealt with the verification of Tn5 transposon mutants. Thirty-two stress sensitive Tn5 transposon mutants affected in 25 genes, which were previously described by Reva *et al.* (2006), were verified for correct transposon insertion sites, and out of these, 15 mutants were selected for further analysis. Growth experiments under stress conditions of high concentrations of sodium benzoate (45 mM) or low temperatures (10°C) were performed to verify stress sensitive phenotypes of the mutants. Besides the cold sensitive mutants, *cbrA*::Tn5, *cbrB*::Tn5, *pcnB*::Tn5, *vacB*::Tn5 and *bipA*::Tn5, which were selected for transcriptome and metabolome analysis, the *pcaI*::Tn5 mutant exhibited strong benzoate sensitivity independent of the concentration. The hypothesis postulated by Reva *et al.* (2006), that there is a bypass in the ß-ketoadipate pathway to compensate for the knock-out in the *pcaIJ* operon caused by the transposon insertion, was disproven in this study. Combinatorial PCR results revealed a partial deletion of the plasposon in *pcaI*::Tn5 cells, which could grow in minimal medium supplemented with benzoate. The deletion included mainly the origin of replication *oriR* of pMB1 (Dennis & Zylstra, 1998). The results demonstrated that the ß-ketoadipate pathway is the only route for benzoate degradation in *P. putida* KT2440. Furthermore, it was shown that plasposons can undergo secondary mutations under high selective pressure, and that partial deletions of the plasposon could rescue *pcaIJ* activity.

For the comprehensive transcriptome analysis of cold shock response of the *P. putida* KT2440 wild type, three transcriptome platforms were applied: Illumina cDNA sequencing and DNA microarrays from Affymetrix and Progenika, respectively. Besides the analysis of the transcriptional profile, the Illumina cDNA results were furthermore used for the transcript analysis of intergenic regions. Since the cDNA sequencing results only reflected the transcriptome status of cells grown under standard condition (M9 + 15 mM succinate) and after cold shock (temperature decrease from 30°C to 10°C), the data were not used for a complete reannotation of the *P. putida* KT2440 genome. However, the results derived from cDNA sequencing revealed a high number of transcripts coding for either proteins or small regulatory RNAs. Out of 143 identified potential hypothetical proteins, the majority showed strong homology to proteins from other *Pseudomonas* species. Another 66 unclassified transcripts remain to be analyzed for being either protein or sRNA coding. Furthermore, 22 sRNAs could be verified in the *P. putida* KT2440 genome due to identified transcript and sequence similarity to already described sRNAs in other Pseudomonads (Livny *et al.*, 2006; Sonnleitner *et al.*, 2008; Gonzaléz *et al.*, 2008; Filiatraut *et al.*, 2010). Among the identified sRNAs, three exhibited differential expression upon cold shock: the TPP riboswitch, sRNA P1 and *crcZ*. The sRNA *crcZ* is located between *cbrB* and *pcnB*, both included in the cold shock experiments, and was about 8-fold down-regulated upon cold shock. The genomic localization of *crcZ* together with the cold sensitive phenotype of Tn5 mutants affected in the flanking genes suggested that the expression level of *crcZ* is involved in adaptation processes to low temperatures. These results demonstrated that many genes have been missed in the initial annotation by Nelson and co-workers (2002), and that many more genes are encoded in the genome than previously thought, of which the majority might be involved in cold shock response, since 87% of the novel hypothetical proteins and 75% of unclassified transcripts were induced at 10°C. The Illumina cDNA sequencing results highlighted the benefits of this new technology in genome-wide transcriptome analysis up to single base-pair resolution and its potential in exploring novel regulatory networks.

The comprehensive transcriptome analysis of *P. putida* wild type revealed 159 genes in total, which were significantly differentially expressed upon cold shock and identified by all three platforms (see

CONCLUSION

Table 3.16). The total number of genes, which were considered to be significantly differentially regulated (fold-change ≥ 2.0, p-value ≤ 0.05, FDR ≤ 0.05 for microarrays or $RPKM_{10°C} > RPKM_{30°C} \pm 3\sqrt{(RPKM_{30°C})}$ for Illumina cDNA sequencing), differed strongly between the three platforms. With cDNA sequencing 2337 genes were found, and with the Affymetrix and Progenika microarrays 1281 and 122 genes, respectively. The low sensitivity of the Progenika microarrays could be ascribed to the design, since only one oligo per ORF was chosen and each oligo was spotted only twice on the arrays. Though the sensitivity was very low with only 2% of all genes (5420 ORFs, Nelson *et al.* (2002)) being significantly differentially expressed, the results of the Progenika microarray analysis were reliable, since the intersection with the other two platforms was high, 104 genes (see Figure 3.22). Though the Spearman correlation of rank number comparison of signal intensities was good (0.74) between the Illumina and Affymetrix platform (see Figure 3.20), the results of platform comparison emphasized the major advantages of cDNA sequencing over genome microarrays in gene expression analysis. Firstly, it displays the real abundance of transcripts, which is not affected by signal intensity saturation and enables furthermore the analysis of low abundance transcripts due to the lower levels of background noise compared to the array technologies. Figure 3.23 highlighted this view. The overall distribution of significantly differentially expressed genes (of Illumina cDNA sequencing and Affymetrix microarray results) in rank number intervals according to Illumina RPKM values showed striking differences. The genes detected with cDNA sequencing were constantly distributed over all intercepts, whereas those detected with Affymetrix microarrays showed a linear decrease corresponding to increasing rank numbers. This demonstrated the limitations of microarrays in detecting low abundance transcripts, which were significantly regulated. Secondly, microarrays are limited in detecting transcripts, which cover the entire genome due to the selection of probes from predefined sequences. The analysis of transcripts in intergenic regions revealed a high number of differentially regulated regions, which were missed with the microarrays.

CONCLUSION

Since the intersection of all three platforms was still good, when the p-value was changed to be significant if, genes with a p-value ≤ 0.1 were considered to be significantly expressed and therefore included in the transcriptome analysis of the wild type cold shock response. In total 159 genes were found by all three transcriptome platforms, and thus were considered to display the key mechanisms in response to a sudden decrease in temperature. Among others, the well characterized chaperones, *groEL* and *groES*, together with their associated trigger factor *tig*, were found to be down-regulated. Considering recent studies (Holtmann *et al.*, 2004; Carruthers & Minion, 2009; Phadtare & Inouye, 2004), these genes are important for adaptation processes to changing temperatures and inhibit either heat shock or cold shock regulatory functions in protein degradation depending on the species. Furthermore, many *algU*-associated genes were found to be differentially regulated. *MucA* and *algN*, which bind to *algU* and thereby repress its expression, were induced upon cold shock. The genes encoding the heat shock proteins DnaK, DnaJ and GrpE, which all have an AlgU-specific promoter site, were repressed. Since membrane fluidity is impaired at low temperatures (Phadtare, 2004), membrane maintenance is a key mechanism in cold shock response, and thus the *algU-mucA-algN* operon known to be involved in cell wall stress (Wood *et al.*, 2006) are also major regulators of cold shock response.

One major finding was that the set of genes repressed upon cold shock was well annotated with only 7% of genes encoding hypothetical proteins, whereas the set of induced genes contained a significantly higher number of genes of unknown function (38%). This showed that much remains to be elucidated in cold shock response of *P. putida*, since the key players in cold shock response are still unknown, which is also supported by the identification of novel transcripts in intergenic regions, which were induced in response to cold shock. For instance, the gene cluster PP1689-PP1691, of which only the first gene is predicted to encode a long-chain fatty acid transporter, is poorly annotated but was found to be highly induced upon cold shock (11-fold, Progenika; 166-fold, Affymetrix; 1232-fold, Illumina). Genes that were repressed upon cold shock mainly reflected a down-regulation of intermediary metabolism with many genes found to be involved in tricarboxylic acid cycle, amino acid metabolism and nucleotide synthesis (see Figure 3.24). Indeed, the metabolism of branched-chain amino acids seemed to be majorly affected.

The proteome analysis revealed 59 proteins to be significantly differentially expressed, of which 18 were only detected by the proteome approach. A high number of membrane-associated proteins (16) was found to be differentially regulated, such as the outer membrane proteins OprQ/H/L/F/I. Another 12 proteins, which were all induced, were associated with transcriptional and translational processes, such as the ribosome-binding factors RimM and RbfA. Both mechanisms were previously described to be affected by cold shock (Phadtare, 2004). However, the proteome and transcriptome data showed a discrepancy in regulation in five out of 41 cases. This indicated that also post-transcriptional mechanisms are involved in adaptation to low temperatures and influence the protein pattern responding to cold shock. By assigning the gene and protein lists to functional categories according to the KEGG annotation of *P. putida* KT2440 (http://www.genome.jp/kegg/), the regulation of translation factors to enhance translation efficiency and the induction of the glycan biosynthesis pathway for the production of cell membrane compounds were identified to represent the first mechanisms in cold shock response. The results showed that the combination of different "omics" is important to enable a complementary view of the regulatory mechanisms. For instance, VacB and BipA were found to be induced on the protein level, but none of the respective genes was identified by the transcriptome approach to be affected upon cold shock.

A comprehensive analysis of transcriptome and metabolome data was applied for the cold shock response of the selected mutants compared to the wild type to reveal the mechanisms that cause a cold sensitive phenotype. The transcriptome profile of the Tn5 mutants analyzed with Progenika microarrays, revealed a consistent down-regulation of a gene cluster, called *ped* cluster in *P. putida* U (Arias *et al.*, 2008), which is independent of the temperature. The *ped* cluster was described to be involved in the conversion of 2-phenylalanine to phenylacetate, which is an intermediate in phenylalanine metabolism. According to the transcriptome and metabolome data of the wild type the metabolism of complex amino acids, reflected by many down-regulated genes involved in the biosynthesis of valine according to the transcriptome data and on the other hand the accumulation of branched-chain and aromatic amino acids according to the metabolome data, is highly affected upon cold shock. Since the Tn5 mutants showed an altered profile of the respective metabolites, mainly reflected by that of *cbrB*::Tn5, suggested that this gene cluster is required for the metabolism of complex amino acids and furthermore other aromatic compounds. A defect in metabolism of branched-chain and aromatic amino acids thus causes a cold sensitive phenotype. The association of the ped cluster with the metabolism of aromatic amino acids and other aromatic compounds was furthermore supported, since all five mutants were also sensitive to high concentration of benzoate, which is structurally related to phenylacetate.

On the one hand, the transcriptome and metabolome data demonstrated the close regulatory connection of the five stress sensitive mutants with the consistent down-regulation of the *ped* cluster and the distinct metabolome profile compared to the wild type. On the other hand, the results supported the highly ranked regulatory role of CbrB, since the transcriptome and metabolome profile were most distinct from the wild type. It is already known that CbrA and CbrB constitute a two-component system, and expression of *pcnB* was recently described to be *cbrB*-dependent

CONCLUSION

(Zhang *et al.*, 2010). The transcriptome profile of the *cbrB*::Tn*5* mutant was the only one that displayed a direct defect in cold shock response. In the *cbrB*-negative background 16 genes belonging to the key players of cold shock response according to the transcriptome data of the wild type were found to be affected in the cold shock response of the mutant. Among others, the expression of PP1689 and PP1690, *bkdA-2*, which is involved in valine biosynthesis, and the two-component system *phoPQ* was found to be *cbrB*-dependent. These genes were induced upon cold shock in the wild type, meaning they are actively required for the stress response, but were repressed in the mutant background. The transcriptome data revealed a *cbrB*-dependent expression of *vacB*. *VacB* was furthermore described to exhibit similar regulatory functions as *crc* (MacGregor *et al.*, 1996), which is also under *cbrB* control (Sonnleitner *et al.*, 2009). In the growth experiments all five mutants exhibited a cold sensitive as well as benzoate sensitive phenotype. However, the results furthermore suggested, since the remaining four mutants did not exhibit such a strong phenotype as the *cbrB*::Tn*5* mutant according to the transcriptome and metabolome data, that a defect in these genes can be somehow compensated. Since *cbrB* is known to be required for a healthy C:N balance, the utilization of certain amino acids (Nishijyo *et al.*, 2001; Zhang & Rainey, 2007) and catabolite repression control (Sonnleitner *et al.*, 2009), the findings of this study emphasizes the global regulatory role of *cbrB* and furthermore demonstrated that it is also required for an intact cold shock response. Together with *cbrA*, *pcnB*, *vacB* and *bipA,* the response regulator CbrB forms a regulatory unit, which links central metabolism, mRNA degradation and translation efficiency. Future experiments will be needed to elucidate the entire regulatory cascade of CbrB.

6 REFERENCES

Aitchison, J. D. and Galitski, T. (2003). Inventories to insights. *J Cell Biol* **161**(3): 465-469.

Albrecht, M.; Sharma, C. M.; Reinhardt, R.; Vogel, J. and Rudel, T. (2010). Deep sequencing-based discovery of the *Chlamydia trachomatis* transcriptome. *Nucleic Acids Res* **38**(3): 868-877.

Altschul, S. F.; Gish, W.; Miller, W.; Myers, E. W. and Lipman, D. J. (1990). Basic local alignment search tool. *J Mol Biol* **215**(3): 403-410.

Anantharaman, V.; Koonin, E. V. and Aravind, L. (2002). Comparative genomics and evolution of proteins involved in RNA metabolism. *Nucleic Acids Res* **30**(7): 1427-1464.

Andrade, J. M.; Cairrão, F. and Arraiano, C. M. (2006). RNase R affects gene expression in stationary phase: regulation of *ompA*. *Mol Microbiol* **60**(1): 219-228.

Ang, D. and Georgopoulos, C. (1989). The heat-shock-regulated *grpE* gene of *Escherichia coli* is required for bacterial growth at all temperatures but is dispensable in certain mutant backgrounds. *J Bacteriol* **171**(5): 2748-2755.

Angelis, M. D. and Gobbetti, M. (2004). Environmental stress responses in *Lactobacillus*: a review. *Proteomics* **4**(1): 106-122.

Aranda-Olmedo, I.; Marín, P.; Ramos, J. L. and Marqués, S. (2006). Role of the *ptsN* gene product in catabolite repression of the *Pseudomonas putida* TOL toluene degradation pathway in chemostat cultures. *Appl Environ Microbiol* **72**(11): 7418-7421.

Arias, S.; Olivera, E. R.; Arcos, M.; Naharro, G. and Luengo, J. M. (2008). Genetic analyses and molecular characterization of the pathways involved in the conversion of 2-phenylethylamine and 2-phenylethanol into phenylacetic acid in *Pseudomonas putida* U. *Environ Microbiol* **10**(2): 413-432.

Ausubel, F. M.; Brent, R.; Kingston, R. E.; Moore, D. D.; Seidman, J. G.; Smith, J. A. and Struhl, K. (1987). *Current Protocols in Molecular Biology*. New York: Greene Publishing Associates and J Wiley and Sons.

Bagdasarian, M.; Lurz, R.; Rückert, B.; Franklin, F. C.; Bagdasarian, M. M.; Frey, J. and Timmis, K. N. (1981). Specific-purpose plasmid cloning vectors. II. Broad host range, high copy number, RSF1010-derived vectors, and a host-vector system for gene cloning in *Pseudomonas*. *Gene* **16**(1-3): 237-247.

Baldwin, B. R.; Mesarch, M. B. and Nies, L. (2000). Broad substrate specificity of naphthalene- and biphenyl-utilizing bacteria. *Appl Microbiol Biotechnol* **53**(6): 748-753.

Bardwell, J. C. and Craig, E. A. (1984). Major heat shock gene of Drosophila and the *Escherichia coli* heat-inducible *dnaK* gene are homologous. *Proc Natl Sci USA* **81**(3): 848-852.

REFERENCES

Barrow, K. and Kwon, D. H. (2009). Alterations in two-component regulatory systems of *phoPQ* and *pmrAB* are associated with polymyxin B resistance in clinical isolates of *Pseudomonas aeruginosa*. *Antimicrob Agents Chemother* **53**(12): 5150-5154.

Behrends, V.; Ryall, B.; Wang, X.; Bundy, J. G. and Williams, H. D. (2010). Metabolic profiling of *Pseudomonas aeruginosa* demonstrates that the anti-sigma factor MucA modulates osmotic stress tolerance. *Mol Biosyst* **6**(3): 562-569.

Bejerano-Sagie, M. and Xavier, K. B. (2007). The role of small RNAs in quorum sensing. *Curr Opin Microbiol* **10**(2): 189-198.

Besemer, J. and Borodovsky, M. (1999). Heuristic approach to deriving models for gene finding. *Nucleic Acids Res* **27**(19): 3911-20.

Birnboim, H. C. and Doly, J. (1979). A rapid alkaline extraction procedure for screening recombinant plasmid DNA. *Nucleic Acids Res* **7**(6): 1513-1523.

Blattner, F. R.; Plunkett, G.; Bloch, C. A.; Perna, N. T.; Burland, V.; Riley, M.; Collado-Vides, J.; Glasner, J. D.; Rode, C. K.; Mayhew, G. F.; Gregor, J.; Davis, N. W.; Kirkpatrick, H. A.; Goeden, M. A.; Rose, D. J.; Mau, B. and Shao, Y. (1997). The complete genome sequence of *Escherichia coli* K-12. *Science* **277**(5331): 1453-1462.

Bloom, J. S.; Khan, Z.; Kruglyak, L.; Singh, M. and Caudy, A. A. (2009). Measuring differential gene expression by short read sequencing: quantitative comparison to 2-channel gene expression microarrays. *BMC Genomics* **10**: 221.

Bollen, A.; Heimark, R. L.; Cozzone, A.; Traut, R. R. and Hershey, J. W. (1975). Cross-linking of initiation factor IF-2 to *Escherichia coli* 30 S ribosomal proteins with dimethylsuberimidate. *J Biol Chem* **250**(11): 4310-4314.

Borowitzka, L. J.; Demmerle, S.; Mackay, M. A. and Norton, R. S. (1980). Carbon-13 Nuclear Magnetic Resonance Study of Osmoregulation in a Blue-Green Alga. *Science* **210**(4470): 650-651.

Bremer, S.; Hoof, T.; Wilke, M.; Busche, R.; Scholte, B.; Riordan, J. R.; Maass, G. and Tümmler, B. (1992). Quantitative expression patterns of multidrug-resistance P-glycoprotein (MDR1) and differentially spliced cystic-fibrosis transmembrane-conductance regulator mRNA transcripts in human epithelia. *Eur J Biochem* **206**(1): 137-149.

Brigulla, M.; Hoffmann, T.; Krisp, A.; Völker, A.; Bremer, E. and Völker, U. (2003). Chill induction of the SigB-dependent general stress response in *Bacillus subtilis* and its contribution to low-temperature adaptation. *J Bacteriol* **185**(15): 4305-4314.

Brown, P. O. and Botstein, D, (1999). Exploring the new world of the genome with DNA microarrays. *Nat Genet* **21**: 33-7.

Bursy, J.; Kuhlmann, A. U.; Pittelkow, M.; Hartmann, H.; Jebbar, M.; Pierik, A. J. and Bremer, E. (2008). Synthesis and uptake of the compatible solutes ectoine and 5-hydroxyectoine by *Streptomyces coelicolor* A3(2) in response to salt and heat stresses. *Appl Environ Microbiol* **74**(23): 7286-7296.

REFERENCES

Cairrão, F.; Cruz, A.; Mori, H. and Arraiano, C. M. (2003). Cold shock induction of RNase R and its role in the maturation of the quality control mediator SsrA/tmRNA. *Mol Microbiol* **50**(4): 1349-1360.

Cao, G. J. and Sarkar, N. (1992). Identification of the gene for an *Escherichia coli* poly(A) polymerase. *Proc Natl Acad Sci U S A* **89**(21): 10380-10384.

Carruthers, M. D. and Minion, C. (2009). Transcriptome analysis of *Escherichia coli* O157:H7 EDL933 during heat shock. *FEMS Microbiol Lett* **295**(1): 96-102.

Charollais, J.; Dreyfus, M. and Iost, I. (2004). CsdA, a cold-shock RNA helicase from *Escherichia coli*, is involved in the biogenesis of 50S ribosomal subunit. *Nucleic Acids Res* **32**(9): 2751-2759.

Cashel, M. and Gallant, J. (1969). Two compounds implicated in the function of the RC gene of *E. coli*. *Nature* **221**: 838–841.

Chen, W. P. and Kuo, T. T. (1993). A simple and rapid method for the preparation of gram negative bacterial genomic DNA. *Nuc Ac Res* **21**(9): 2260.

Chen, C. Y. and Morse, S. A. (1999). *Neisseria gonorrhoeae* bacterioferritin: structural heterogeneity, involvement in iron storage and protection against oxidative stress. *Microbiology* **145** (Pt 10): 2967-2975.

Cheng, Z. F.; Zuo, Y.; Li, Z.; Rudd, K. E. and Deutscher, M. P. (1998). The *vacB* gene required for virulence in *Shigella flexneri* and *Escherichia coli* encodes the exoribonuclease RNase R. *J Biol Chem* **273**(23): 14077-14080.

Cheng, Z.-F. and Deutscher, M. P. (2003). Quality control of ribosomal RNA mediated by polynucleotide phosphorylase and RNase R. *Proc Natl Sci USA* **100**(11): 6388-6393.

Cheng, Z.-F. and Deutscher, M. P. (2005). An important role for RNase R in mRNA decay. *Mol Cell* **17**(2): 313-318.

Cole, S. T.; Brosch, R.; Parkhill, J.; Garnier, T.; Churcher, C.; Harris, D.; Gordon, S. V.; Eiglmeier, K.; Gas, S.; Barry, C. E.; Tekaia, F.; Badcock, K.; Basham, D.; Brown, D.; Chillingworth, T.; Connor, R.; Davies, R.; Devlin, K.; Feltwell, T.; Gentles, S.; Hamlin, N.; Holroyd, S.; Hornsby, T.; Jagels, K.; Krogh, A.; McLean, J.; Moule, S.; Murphy, L.; Oliver, K.; Osborne, J.; Quail, M. A.; Rajandream, M. A.; Rogers, J.; Rutter, S.; Seeger, K.; Skelton, J.; Squares, R.; Squares, S.; Sulston, J. E.; Taylor, K.; Whitehead, S. and Barrell, B. G. (1998). Deciphering the biology of *Mycobacterium tuberculosis* from the complete genome sequence. *Nature* **393**(6685): 537-544.

Comes, J. E. and Beelman, R. B. (2002). Addition of fumaric acid and sodium benzoate as an alternative method to achieve a 5-log reduction of *Escherichia coli* O157:H7 populations in apple cider. *J Food Prot* **65**(3): 476-483.

Confalonieri, F. and Duguet, M. (1995). A 200-amino acid ATPase module in search of a basic function. *Bioessays* **17**(7): 639-650.

REFERENCES

Cowles, C. E.; Nichols, N. N. and Harwood, C. S. (2000). BenR, a XylS homologue, regulates three different pathways of aromatic acid degradation in *Pseudomonas putida*. *J Bacteriol* **182**(22): 6339-6346.

Daran-Lapujade P.; Daran J.-M.;. van Maris, A. J. A.; de Winde, J. H. and Pronk J. T. (2008). Chemostat-Based Micro-Array Analysis in Baker's Yeast. *Advances in Microbial Physiology* **54**: 257-311.

Davenport, C. F.; Wiehlmann, L.; Reva, O. N. and Tümmler, B. (2009). Visualization of *Pseudomonas* genomic structure by abundant 8-14mer oligonucleotides. *Environ Microbiol* **11**(5): 1092-1104.

Delcher, A. L.; Harmon, D.; Kasif, S.; White, O. and Salzberg, S. L. (1999). Improved microbial gene identification with GLIMMER. *Nucleic Acids Res* **27**(23): 4636-41.

DeLivron, M. A. and Robinson, V. L. (2008). *Salmonella enterica* serovar Typhimurium BipA exhibits two distinct ribosome binding modes. *J Bacteriol* **190**(17): 5944-5952.

DeLivron, M. A.; Makanji, H. S.; Lane, M. C. and Robinson, V. L. (2009). A novel domain in translational GTPase BipA mediates interaction with the 70S ribosome and influences GTP hydrolysis. *Biochemistry* **48**(44): 10533-10541.

de Lorenzo, V.; Herrero, M.; Jakubzik, U. and Timmis, K. N. (1990). Mini-Tn5 transposon derivatives for insertion mutagenesis, promoter probing, and chromosomal insertion of cloned DNA in gram-negative eubacteria. *J Bacteriol* **172**(11): 6568-6572.

de Lorenzo, V. and Timmis, K. N. (1994). Analysis and construction of stable phenotypes in gram-negative bacteria with Tn5- and Tn10-derived minitransposons. *Methods Enzymol* **235**: 386-405.

Dennis, J. J. and Zylstra, G. J. (1998). Plasposons: modular self-cloning minitransposon derivatives for rapid genetic analysis of gram-negative bacterial genomes. *Appl Environ Microbiol* **64**(7): 2710-2715.

Dubern, J.-F.; Lagendijk, E. L.; Lugtenberg, B. J. J. and Bloemberg, G. V. (2005). The heat shock genes *dnaK*, *dnaJ*, and *grpE* are involved in regulation of putisolvin biosynthesis in *Pseudomonas putida* PCL1445. *J Bacteriol* **187**(17): 5967-5976.

Dunn, H. D.; Curtin, T.; Oriordan, M. A.; Coen, P.; Kieran, P. M.; Malone, D. M. and OConnor, K. E. (2005). Aromatic and aliphatic hydrocarbon consumption and transformation by the styrene degrading strain *Pseudomonas putida* CA-3. *FEMS Microbiol Lett* **249**(2): 267-273.

Dunn, M. F. (1998). Tricarboxylic acid cycle and anaplerotic enzymes in rhizobia. *FEMS Microbiol Rev* **22**(2): 105-123.

Duo, M.; Hou, S. and Ren, D. (2008). Identifying *Escherichia coli* genes involved in intrinsic multidrug resistance. *Appl Microbiol Biotechnol* **81**(4): 731-741.

REFERENCES

Duque, E.; Segura, A.; Mosqueda, G. and Ramos, J. L. (2001). Global and cognate regulators control the expression of the organic solvent efflux pumps TtgABC and TtgDEF of *Pseudomonas putida*. *Mol Microbiol* **39**(4): 1100-1106.

Elles, L. M. S.; Sykes, M. T.; Williamson, J. R. and Uhlenbeck, O. C. (2009). A dominant negative mutant of the *E. coli* RNA helicase DbpA blocks assembly of the 50S ribosomal subunit. *Nucleic Acids Res* **37**(19): 6503-6514.

Enderle, P. J. and Farwell, M. A. (1998). Electroporation of freshly plated *Escherichia coli* and *Pseudomonas aeruginosa* cells. *Biotechniques* **25**(6): 954-6, 958.

Erova, T. E.; Kosykh, V. G.; Fadl, A. A.; Sha, J.; Horneman, A. J. and Chopra, A. K. (2008). Cold shock exoribonuclease R (VacB) is involved in *Aeromonas hydrophila* pathogenesis. *J Bacteriol* **190**(10): 3467-3474.

Expert, D.; Sauvage, C. and Neilands, J. B. (1992). Negative transcriptional control of iron transport in *Erwinia chrysanthemi* involves an iron-responsive two-factor system. *Mol Microbiol* **6**(14): 2009-2017.

Farris, M.; Grant, A.; Richardson, T. B. and OConnor, C. D. (1998). BipA: a tyrosine-phosphorylated GTPase that mediates interactions between enteropathogenic *Escherichia coli* (EPEC) and epithelial cells. *Mol Microbiol* **28**(2): 265-279.

Feist, C. F. and Hegeman, G. D. (1969a). Regulation of the meta cleavage pathway for benzoate oxidation by *Pseudomonas putida*. *J Bacteriol* **100**(2): 1121-1123.

Feist, C. F. and Hegeman, G. D. (1969b). Phenol and benzoate metabolism by *Pseudomonas putida*: regulation of tangential pathways. *J Bacteriol* **100**(2): 869-877.

Filiatrault, M. J.; Stodghill, P. V.; Bronstein, P. A.; Moll, S.; Lindeberg, M.; Grills, G.; Schweitzer, P.; Wang, W.; Schroth, G. P.; Luo, S.; Khrebtukova, I.; Yang, Y.; Thannhauser, T.; Butcher, B. G.; Cartinhour, S. and Schneider, D. J. (2010). Transcriptome analysis of *Pseudomonas syringae* identifies new genes, ncRNAs, and antisense activity. *J Bacteriol* **192**(9): 2359-72.

Finn, R. D.; Tate, J.; Mistry, J.; Coggill, P. C.; Sammut, S. J.; Hotz, H.-R.; Ceric, G.; Forslund, K.; Eddy, S. R.; Sonnhammer, E. L. L. and Bateman, A. (2008). The Pfam protein families database. *Nucleic Acids Res* **36**(Database issue): D281-D288.

Firoved, A. M.; Boucher, J. C. and Deretic, V. (2002). Global genomic analysis of AlgU (sigma(E))-dependent promoters (sigmulon) in *Pseudomonas aeruginosa* and implications for inflammatory processes in cystic fibrosis. *J Bacteriol* **184**(4): 1057-1064.

Fleischmann, R. D.; Adams, M. D.; White, O.; Clayton, R. A.; Kirkness, E. F.; Kerlavage, A. R.; Bult, C. J.; Tomb, J. F.; Dougherty, B. A. and Merrick, J. M. (1995). Whole-genome random sequencing and assembly of *Haemophilus influenzae* Rd. *Science* **269**(5223): 496-512.

Fozo, E. M.; Hemm, M. R. and Storz, G. (2008). Small toxic proteins and the antisense RNAs that repress them. *Microbiol Mol Biol Rev* **72**(4): 579-89.

REFERENCES

Frimmersdorf, E.; Horatzek, S.; Pelnikevich, A.; Wiehlmann, L. and Schomburg, D. (2010). How *Pseudomonas aeruginosa* adapts to various environments: a metabolic approach. *Env Microbiol* (in press).

Fromont-Racine, M.; Mayes, A. E.; Brunet-Simon, A.; Rain, J. C.; Colley, A.; Dix, I.; Decourty, L.; Joly, N.; Ricard, F.; Beggs, J. D. and Legrain, P. (2000). Genome-wide protein interaction screens reveal functional networks involving Sm-like proteins. *Yeast* 17(2): 95-110.

Fromont-Racine, M.; Rain, J.-C. and Legrain, P. (2002). Building protein-protein networks by two-hybrid mating strategy. *Methods Enzymol* 350: 513-524.

Ganesan, H.; Rakitianskaia, A. S.; Davenport, C. F.; Tümmler, B. and Reva, O. N. (2008). The SeqWord Genome Browser: an online tool for the identification and visualization of atypical regions of bacterial genomes through oligonucleotide usage. *BMC Bioinformatics* 9: 333.

García-Estepa, R.; Argandoña, M.; Reina-Bueno, M.; Capote, N.; Iglesias-Guerra, F.; Nieto, J. J. and Vargas, C. (2006). The *ectD* gene, which is involved in the synthesis of the compatible solute hydroxyectoine, is essential for thermoprotection of the halophilic bacterium *Chromohalobacter salexigens*. *J Bacteriol* 188(11): 3774-3784.

Gardy, J. L.; Laird, M. R.; Chen, F.; Rey, S.; Walsh, C. J.; Ester, M. and Brinkman, F. S. (2005). PSORTb v.2.0: expanded prediction of bacterial protein subcellular localization and insights gained from comparative proteome analysis. *Bioinformatics* 21(5): 617-23.

Garrity-Ryan, L.; Kazmierczak, B.; Kowal, R.; Comolli, J.; Hauser, A. and Engel, J. N. (2000). The arginine finger domain of ExoT contributes to actin cytoskeleton disruption and inhibition of internalization of *Pseudomonas aeruginosa* by epithelial cells and macrophages. *Infect Immun* 68(12): 7100-7113.

Gattiker, A.; Michoud, K.; Rivoire, C.; Auchincloss, A. H.; Coudert, E.; Lima, T.; Kersey, P.; Pagni, M.; Sigrist, C. J.; Lachaize, C.; Veuthey, A. L.; Gasteiger, E. and Bairoch, A. (2003). Automated annotation of microbial proteomes in SWISS-PROT. *Comput Biol Chem* 27(1): 49-58.

Gautier, L.; Cope, L.; Bolstad, B. M. and Irizarry, R. A. (2004). affy-analysis of Affymetrix GeneChip data at the probe level. *Bioinformatics* 20(3): 307-315.

Giardine, B.; Riemer, C.; Hardison, R. C.; Burhans, R.; Elnitski, L.; Shah, P.; Zhang, Y.; Blankenberg, D.; Albert, I.; Taylor, J.; Miller, W.; Kent, W. J. and Nekrutenko, A. (2005). Galaxy: a platform for interactive large-scale genome analysis. *Genome Res* 15(10): 1451-1455.

Gibson, K. E.; Campbell, G. R.; Lloret, J. and Walker, G. C. (2006). CbrA is a stationary-phase regulator of cell surface physiology and legume symbiosis in *Sinorhizobium meliloti*. *J Bacteriol* 188(12): 4508-4521.

Gibson, K. E.; Barnett, M. J.; Toman, C. J.; Long, S. R. and Walker, G. C. (2007). The symbiosis regulator CbrA modulates a complex regulatory network affecting the flagellar apparatus and cell envelope proteins. *J Bacteriol* 189(9): 3591-3602.

REFERENCES

Giuliodori, A. M.; Pietro, F. D.; Marzi, S.; Masquida, B.; Wagner, R.; Romby, P.; Gualerzi, C. O. and Pon, C. L. (2010). The *cspA* mRNA is a thermosensor that modulates translation of the cold-shock protein CspA. *Mol Cell* **37**(1): 21-33.

Goff, M.; Nikodinovic-Runic, J. and OConnor, K. E. (2009). Characterization of temperature-sensitive and lipopolysaccharide overproducing transposon mutants of *Pseudomonas putida* CA-3 affected in PHA accumulation. *FEMS Microbiol Lett* **292**(2): 297-305.

Goldberg, J. B.; Won, J. and Ohman, D. E. (1990). Precise excision and instability of the transposon Tn*5* in *Pseudomonas aeruginosa*. *J Gen Microbiol* **136**(5): 789-796.

González, N.; Heeb, S.; Valverde, C.; Kay, E.; Reimmann, C.; Junier, T. and Haas, D. (2008). Genome-wide search reveals a novel GacA-regulated small RNA in *Pseudomonas* species. *BMC Genomics* **9**: 167.

Gooderham, W. J.; Gellatly, S. L.; Sanschagrin, F.; McPhee, J. B.; Bains, M.; Cosseau, C.; Levesque, R. C. and Hancock, R. E. W. (2009). The sensor kinase PhoQ mediates virulence in *Pseudomonas aeruginosa*. *Microbiology* **155**(Pt 3): 699-711.

Govan, J. R. and Deretic, V. (1996). Microbial pathogenesis in cystic fibrosis: mucoid *Pseudomonas aeruginosa* and *Burkholderia cepacia*. *Microbiol Rev* **60**(3): 539-574.

Grant, A. J.; Farris, M.; Alefounder, P.; Williams, P. H.; Woodward, M. J. and OConnor, C. D. (2003). Co-ordination of pathogenicity island expression by the BipA GTPase in enteropathogenic *Escherichia coli* (EPEC). *Mol Microbiol* **48**(2): 507-521.

Greated, A.; Lambertsen, L.; Williams, P. A. and Thomas, C. M. (2002). Complete sequence of the IncP-9 TOL plasmid pWW0 from *Pseudomonas putida*. *Environ Microbiol* **4**(12): 856-871.

Griffin, T. J. and Aebersold, R. (2001). Advances in proteome analysis by mass spectrometry. *J Biol Chem* **276**(49): 45497-45500.

Gross, F.; Gottschalk, D. and Müller, R. (2005). Posttranslational modification of myxobacterial carrier protein domains in *Pseudomonas* sp. by an intrinsic phosphopantetheinyl transferase. *Appl Microbiol Biotechnol* **68**(1): 66-74.

Gross, F.; Luniak, N.; Perlova, O.; Gaitatzis, N.; Jenke-Kodama, H.; Gerth, K.; Gottschalk, D.; Dittmann, E. and Müller, R. (2006a). Bacterial type III polyketide synthases: phylogenetic analysis and potential for the production of novel secondary metabolites by heterologous expression in *pseudomonads*. *Arch Microbiol* **185**(1): 28-38.

Gross, F.; Ring, M. W.; Perlova, O.; Fu, J.; Schneider, S.; Gerth, K.; Kuhlmann, S.; Stewart, A. F.; Zhang, Y. and Müller, R. (2006b). Metabolic engineering of *Pseudomonas putida* for methylmalonyl-CoA biosynthesis to enable complex heterologous secondary metabolite formation. *Chem Biol* **13**(12): 1253-1264.

Gryder, R. M. and Adams, E. (1969). Inducible degradation of hydroxyproline in *Pseudomonas putida*: pathway regulation and hydroxyproline uptake. *J Bacteriol* **97**(1): 292-306.

REFERENCES

Gryder, R. M. and Adams, E. (1970). Properties of the inducible hydroxyproline transport system of *Pseudomonas putida*. *J Bacteriol* **101**(3):48-58.

Gualerzi, C. O.; Giuliodori, A. M. and Pon, C. L. (2003). Transcriptional and post-transcriptional control of cold-shock genes. *J Mol Biol* **331**(3): 527-539.

Guiliano, D.; Ganatra, M.; Ware, J.; Parrot, J.; Daub, J.; Moran, L.; Brennecke, H.; Foster, J. M.; Supali, T.; Blaxter, M.; Scott, A. L.; Williams, S. A. and Slatko, B. E. (1999). Chemiluminescent detection of sequential DNA hybridizations to high-density, filter-arrayed cDNA libraries: a subtraction method for novel gene discovery. *Biotechniques* **27**(1):146-52.

Hagemann, M.; Richter, S.; Zuther, E. and Schoor, A. (1996). Characterization of a glucosylglycerol-phosphate-accumulating, salt-sensitive mutant of the cyanobacterium *Synechocystis* sp. strain PCC 6803. *Arch Microbiol* **166**(2): 83-91.

Hagemann, M.; Richter, S. and Mikkat, S. (1997). The *ggtA* gene encodes a subunit of the transport system for the osmoprotective compound glucosylglycerol in *Synechocystis* sp. strain PCC 6803. *J Bacteriol* **179**(3): 714-720.

Hall, N. (2007). Advanced sequencing technologies and their wider impact in microbiology. *J Exp Biol* **210**(Pt 9): 1518-1525.

Hallsworth, J. E.; Heim, S. and Timmis, K. N. (2003). Chaotropic solutes cause water stress in *Pseudomonas putida*. *Environ Microbiol* **5**(12): 1270-1280.

Harishchandra, R. K.; Wulff, S.; Lentzen, G.; Neuhaus, T. and Galla, H.-J. (2010). The effect of compatible solute ectoines on the structural organization of lipid monolayer and bilayer membranes. *Biophys Chem*. (in print).

Harwood, C. S. and Parales, R. E. (1996). The beta-ketoadipate pathway and the biology of self-identity. *Annu Rev Microbiol* **50**: 553-590.

Haseltine, W. A. and Block, R. (1973). Synthesis of guanosine tetra- and pentaphosphate requires the presence of a codon-specific, uncharged transfer ribonucleic acid in the acceptor site of ribosomes. *Proc Natl Sci USA* **70**(5): 1564-1568.

Hazelbauer, G. L. and Harayama, S. (1983). Sensory transduction in bacterial chemotaxis. *Int Rev Cytol* **81**: 33-70.

Hecker, M. and Völker, U. (2001). General stress response of *Bacillus subtilis* and other bacteria. *Adv Microb Physiol* **44**: 35-91.

Hemm, M. R.; Paul, B. J.; Schneider, T. D.; Storz, G. and Rudd, K. E. (2008). Small membrane proteins found by comparative genomics and ribosome binding site models. *Mol Microbiol* **70**(6): 1487-1501.

Hemm, M. R.; Paul, B. J.; Miranda-Ríos, J.; Zhang, A.; Soltanzad, N. and Storz, G. (2010). Small stress response proteins in *Escherichia coli*: proteins missed by classical proteomic studies. *J Bacteriol* **192**(1): 46-58.

REFERENCES

Hensel, M.; Shea, J. E.; Gleeson, C.; Jones, M. D.; Dalton, E. and Holden, D. W. (1995). Simultaneous identification of bacterial virulence genes by negative selection. *Science* 269(5222): 400-403.

Henzel, W. J.; Billeci, T. M.; Stults, J. T.; Wong, S. C.; Grimley, C. and Watanabe, C. (1993). Identifying proteins from two-dimensional gels by molecular mass searching of peptide fragments in protein sequence databases. *Proc Natl Acad Sci U S A* 90(11): 5011-5015.

Herrero, M.; de Lorenzo, V. and Timmis, K. N. (1990). Transposon vectors containing non-antibiotic resistance selection markers for cloning and stable chromosomal insertion of foreign genes in gram-negative bacteria. *J Bacteriol* 172(11): 6557-6567.

Hester, K. L.; Lehman, J.; Najar, F.; Song, L.; Roe, B. A.; MacGregor, C. H.; Hager, P. W.; Phibbs, P. V. and Sokatch, J. R. (2000). Crc is involved in catabolite repression control of the *bkd* operons of *Pseudomonas putida* and *Pseudomonas aeruginosa*. *J Bacteriol* 182(4): 1144-1149.

Hobbs, E. C.; Astarita, J. L. and Storz, G. (2010). Small RNAs and small proteins involved in resistance to cell envelope stress and acid shock in *Escherichia coli*: analysis of a bar-coded mutant collection. *J Bacteriol* 192(1): 59-67.

Hoch, J. A. and Silhavy, T. J. (1995). *Two-component signal transduction*. Washington, DC: Am. Soc. Microbiol.

't Hoen, P. A. C.; Ariyurek, Y.; Thygesen, H. H.; Vreugdenhil, E.; Vossen, R. H. A. M.; de Menezes, R. X.; Boer, J. M.; van Ommen, G.-J. B. and den Dunnen, J. T. (2008). Deep sequencing-based expression analysis shows major advances in robustness, resolution and inter-lab portability over five microarray platforms. *Nucleic Acids Res* 36(21): e141.

Holtmann, G.; Brigulla, M.; Steil, L.; Schütz, A.; Barnekow, K.; Völker, U. and Bremer, E. (2004). RsbV-independent induction of the SigB-dependent general stress regulon of *Bacillus subtilis* during growth at high temperature. *J Bacteriol* 186(18): 6150-6158.

Hong, S. J.; Tran, Q. A. and Keiler, K. C. (2005). Cell cycle-regulated degradation of tmRNA is controlled by RNase R and SmpB. *Mol Microbiol* 57(2): 565-75.

Hoshino, T.; Kose-Terai, K. and Sato, K. (1992). Solubilization and reconstitution of the *Pseudomonas aeruginosa* high affinity branched-chain amino acid transport system. *J Biol Chem* 267(30): 21313-21318.

Hunger, K.; Beckering, C. L.; Wiegeshoff, F.; Graumann, P. L. and Marahiel, M. A. (2006). Cold-induced putative DEAD box RNA helicases CshA and CshB are essential for cold adaptation and interact with cold shock protein B in *Bacillus subtilis*. *J Bacteriol* 188(1): 240-248.

Inoue, A.; Yamamoto, M. and Horikoshi, K. (1991). *Pseudomonas putida* Which Can Grow in the Presence of Toluene. *Appl Environ Microbiol* 57(5): 1560-1562.

Jacobs, M. A.; Alwood, A.; Thaipisuttikul, I.; Spencer, D.; Haugen, E.; Ernst, S.; Will, O.; Kaul, R.; Raymond, C.; Levy, R.; Chun-Rong, L.; Guenthner, D.; Bovee, D.; Olson, M. V. and Manoil, C. (2003). Comprehensive transposon mutant library of *Pseudomonas aeruginosa*. *Proc Natl Sci USA* **100**(24): 14339-14344.

Jain, C. (2008). The *E. coli* RhlE RNA helicase regulates the function of related RNA helicases during ribosome assembly. *RNA* **14**(2): 381-389.

Janke, D.; Pohl, R. and Fritsche, W. (1981). Regulation of phenol degradation in *Pseudomonas putida*. *Z Allg Mikrobiol* **21**(4): 295-303.

Jäger, D.; Sharma, C. M.; Thomsen, J.; Ehlers, C.; Vogel, J. and Schmitz, R. A. (2009). Deep sequencing analysis of the *Methanosarcina mazei* Göl transcriptome in response to nitrogen availability. *Proc Natl Acad Sci U S A* **106**(51): 21878-21882.

Jensen, L. B.; Ramos, J. L.; Kaneva, Z. and Molin, S. (1993). A substrate-dependent biological containment system for *Pseudomonas putida* based on the *Escherichia coli gef* gene. *Appl Environ Microbiol* **59**(11): 3713-3717.

Kandror, O.; Sherman, M.; Rhode, M. and Goldberg, A. L. (1995). Trigger factor is involved in GroEL-dependent protein degradation in *Escherichia coli* and promotes binding of GroEL to unfolded proteins. *EMBO J* **14**(23): 6021-6027.

Kandror, O. and Goldberg, A. L. (1997). Trigger factor is induced upon cold shock and enhances viability of *Escherichia coli* at low temperatures. *Proc Natl Acad Sci U S A* **94**(10): 4978-4981.

Kandror, O.; Sherman, M.; Moerschell, R. and Goldberg, A. L. (1997). Trigger factor associates with GroEL in vivo and promotes its binding to certain polypeptides. *J Biol Chem* **272**(3): 1730-1734.

Kanehisa, M.; Araki, M.; Goto, S.; Hattori, M.; Hirakawa, M.; Itoh, M.; Katayama, T.; Kawashima, S.; Okuda, S.; Tokimatsu, T. and Yamanishi, Y. (2008). KEGG for linking genomes to life and the environment. *Nucleic Acids Res* **36**(Database issue): D480-D484.

Keith, L. M.; Partridge, J. E. and Bender, C. L. (1999). *dnaK* and the heat stress response of *Pseudomonas syringae* pv. *glycinea*. *Mol Plant Microbe Interact* **12**(7): 563-574.

Kim, K. K.; Kim, R. and Kim, S. H. (1998). Crystal structure of a small heat-shock protein. *Nature* **394**(6693): 595-599.

Kim, S. K.; Kimura, S.; Shinagawa, H.; Nakata, A.; Lee, K. S.; Wanner, B. L. and Makino, K. (2000). Dual transcriptional regulation of the *Escherichia coli* phosphate-starvation-inducible *psiE* gene of the phosphate regulon by PhoB and the cyclic AMP (cAMP)-cAMP receptor protein complex. *J Bacteriol* **182**(19): 5596-5599.

Kiss, E.; Huguet, T.; Poinsot, V. and Batut, J. (2004). The *typA* gene is required for stress adaptation as well as for symbiosis of *Sinorhizobium meliloti* 1021 with certain *Medicago truncatula* lines. *Mol Plant Microbe Interact* **17**(3): 235-244.

REFERENCES

Kleerebezem, M. (2004). Quorum sensing control of lantibiotic production; nisin and subtilin autoregulate their own biosynthesis. *Peptides* **25**(9): 1405-1414.

Klein, W.; Weber, M.H. and Marahiel, M. A. (1999). Cold shock response of Bacillus subtilis: isoleucine-dependent switch in the fatty acid branching pattern for membrane adaptation to low temperatures. *J Bacteriol* **181**(17): 5341-9.

Kleppe, K.; Ohtsuka, E.; Kleppe, R.; Molineux, I. and Khorana, H. G. (1971). Studies on polynucleotides. XCVI. Repair replications of short synthetic DNAs as catalyzed by DNA polymerases. *J Mol Biol* **56**(2): 341-361.

Klockgether, J.; Munder, A.; Neugebauer, J.; Davenport, C. F.; Stanke, F.; Larbig, K. D.; Heeb, S.; Schöck, U.; Pohl, T. M.; Wiehlmann, L. and Tümmler, B. (2010). Genome diversity of *Pseudomonas aeruginosa* PAO1 laboratory strains. *J Bacteriol* **192**(4): 1113-1121.

Ko, R.; Smith, L. T. and Smith, G. M. (1994). Glycine betaine confers enhanced osmotolerance and cryotolerance on *Listeria monocytogenes*. *J Bacteriol* **176**(2): 426-431.

Kozarewa, I.; Ning, Z.; Quail, M. A.; Sanders, M. J.; Berriman, M. and Turner, D. J. (2009). Amplification-free Illumina sequencing-library preparation facilitates improved mapping and assembly of (G+C)-biased genomes. *Nat Methods* **6**(4): 291-295.

Krishnan, K. and Flower, A. M. (2008). Suppression of Delta*bipA* phenotypes in *Escherichia coli* by abolishment of pseudouridylation at specific sites on the 23S rRNA. *J Bacteriol* **190**(23): 7675-7683.

Krogh, A.; Larsson, B.; von Heijne, G. and Sonnhammer, E. L. (2001). Predicting transmembrane protein topology with a hidden Markov model: application to complete genomes. *J Mol Biol* **305**(3): 567-580.

Kwon, Y. M. and Ricke, S. C. (2000). Efficient amplification of multiple transposon-flanking sequences. *J Microbiol Methods* **41**(3): 195-199.

Kwon, Y. T.; Lévy F. and Varshavsky, A. (1999). Bivalent inhibitor of the N-end rule pathway. *J Biol Chem* **274**(25): 18135-9.

Langmead, B.; Trapnell, C.; Pop, M. and Salzberg, S. L. (2009). Ultrafast and memory-efficient alignment of short DNA sequences to the human genome. *Genome Biol* **10**(3): R25.

Lapouge, K.; Sineva, E.; Lindell, M.; Starke, K.; Baker, C. S.; Babitzke, P. and Haas, D. (2007). Mechanism of *hcnA* mRNA recognition in the Gac/Rsm signal transduction pathway of *Pseudomonas fluorescens*. *Mol Microbiol* **66**(2): 341-356.

Lapouge, K.; Schubert, M.; Allain, F. H.-T. and Haas, D. (2008). Gac/Rsm signal transduction pathway of gamma-proteobacteria: from RNA recognition to regulation of social behaviour. *Mol Microbiol* **67**(2): 241-253.

Lee, S.; Sowa, M. E.; hei Watanabe, Y.; Sigler, P. B.; Chiu, W.; Yoshida, M. and Tsai, F. T. F. (2003). The structure of ClpB: a molecular chaperone that rescues proteins from an aggregated state. *Cell* **115**(2): 229-240.

Li, W. and Lu, C.-D. (2007). Regulation of carbon and nitrogen utilization by CbrAB and NtrBC two-component systems in *Pseudomonas aeruginosa*. *J Bacteriol* **189**(15): 5413-5420.

Li, Z.; Reimers, S.; Pandit, S. and Deutscher, M. P. (2002). RNA quality control: degradation of defective transfer RNA. *EMBO J* **21**(5): 1132-1138.

Liberati, N. T.; Urbach, J. M.; Miyata, S.; Lee, D. G.; Drenkard, E.; Wu, G.; Villanueva, J.; Wei, T. and Ausubel, F. M. (2006). An ordered, nonredundant library of *Pseudomonas aeruginosa* strain PA14 transposon insertion mutants. *Proc Natl Acad Sci U S A* **103**(8): 2833-2838.

Liberati, N. T.; Urbach, J. M.; Thurber, T. K.; Wu, G. and Ausubel, F. M. (2008). Comparing insertion libraries in two *Pseudomonas aeruginosa* strains to assess gene essentiality. *Methods Mol Biol* **416**: 153-169.

Lindquist, S. and Craig, E. A. (1988). The heat-shock proteins. *Annu Rev Genet* **22**: 631-677.

Liu, J. D. and Parkinson, J. S. (1989). Genetics and sequence analysis of the *pcn*B locus, an *Escherichia coli* gene involved in plasmid copy number control. *J Bacteriol* **171**(3): 1254-1261.

Livny, J.; Brencic, A.; Lory, S. and Waldor, M. K. (2006). Identification of 17 *Pseudomonas aeruginosa* sRNAs and prediction of sRNA-encoding genes in 10 diverse pathogens using the bioinformatic tool sRNAPredict2. *Nucleic Acids Res* **34**(12): 3484-3493.

Lockwood, A. H.; Sarkar, P. and Maitra, U. (1972). Release of polypeptide chain initiation factor IF-2 during initiation complex formation. *Proc Natl Acad Sci U S A* **69**(12): 3602-3605.

Lopilato, J.; Bortner, S. and Beckwith, J. (1986). Mutations in a new chromosomal gene of *Escherichia coli* K-12, *pcnB*, reduce plasmid copy number of pBR322 and its derivatives. *Mol Gen Genet* **205**(2): 285-290.

Lu, J.; Aoki, H. and Ganoza, M. C. (1999). Molecular characterization of a prokaryotic translation factor homologous to the eukaryotic initiation factor eIF4A. *Int J Biochem Cell Biol* **31**(1): 215-29.

Luengo, J. M.; García, J. L. and Olivera, E. R. (2001). The phenylacetyl-CoA catabolon: a complex catabolic unit with broad biotechnological applications. *Mol Microbiol* **39**(6): 1434-1442.

Lukashin, A. V. and Borodovsky, M. (1998). GeneMark.hmm: new solutions for gene finding. *Nucleic Acids Res* **26**(4): 1107-1115.

Macfarlane, E. L.; Kwasnicka, A.; Ochs, M. M. and Hancock, R. E. (1999). PhoP-PhoQ homologues in *Pseudomonas aeruginosa* regulate expression of the outer-membrane protein OprH and polymyxin B resistance. *Mol Microbiol* **34**(2): 305-316.

Macfarlane, E. L.; Kwasnicka, A. and Hancock, R. E. (2000). Role of *Pseudomonas aeruginosa* PhoP-phoQ in resistance to antimicrobial cationic peptides and aminoglycosides. *Microbiology* **146**: 2543-2554.

REFERENCES

MacGregor, C. H.; Arora, S. K.; Hager, P. W.; Dail, M. B. and Phibbs, P. V. (1996). The nucleotide sequence of the *Pseudomonas aeruginosa pyrE-crc-rph* region and the purification of the *crc* gene product. *J Bacteriol* **178**(19): 5627-5635.

Madhusudhan, K. T.; Lorenz, D. and Sokatch, J. R. (1993). The *bkdR* gene of *Pseudomonas putida* is required for expression of the *bkd* operon and encodes a protein related to Lrp of *Escherichia coli*. *J Bacteriol* **175**(13): 3934-3940.

Madhusudhan, K. T.; Hester, K. L.; Friend, V. and Sokatch, J. R. (1997). Transcriptional activation of the *bkd* operon of *Pseudomonas putida* by BkdR. *J Bacteriol* **179**(6): 1992-1997.

Margulies, M.; Egholm, M.; Altman, W. E.; Attiya, S.; Bader, J. S.; Bemben, L. A.; Berka, J.; Braverman, M. S.; Chen, Y.-J.; Chen, Z.; Dewell, S. B.; Du, L.; Fierro, J. M.; Gomes, X. V.; Godwin, B. C.; He, W.; Helgesen, S.; Ho, C. H.; Ho, C. H.; Irzyk, G. P.; Jando, S. C.; Alenquer, M. L. I.; Jarvie, T. P.; Jirage, K. B.; Kim, J.-B.; Knight, J. R.; Lanza, J. R.; Leamon, J. H.; Lefkowitz, S. M.; Lei, M.; Li, J.; Lohman, K. L.; Lu, H.; Makhijani, V. B.; McDade, K. E.; McKenna, M. P.; Myers, E. W.; Nickerson, E.; Nobile, J. R.; Plant, R.; Puc, B. P.; Ronan, M. T.; Roth, G. T.; Sarkis, G. J.; Simons, J. F.; Simpson, J. W.; Srinivasan, M.; Tartaro, K. R.; Tomasz, A.; Vogt, K. A.; Volkmer, G. A.; Wang, S. H.; Wang, Y.; Weiner, M. P.; Yu, P.; Begley, R. F. and Rothberg, J. M. (2005). Genome sequencing in microfabricated high-density picolitre reactors. *Nature* **437**(7057): 376-380.

Marioni, J. C.; Mason, C. E.; Mane, S. M.; Stephens, M. and Gilad, Y. (2008). RNA-seq: an assessment of technical reproducibility and comparison with gene expression arrays. *Genome Res* **18**(9): 1509-1517.

Marmur, J. and Doty, P. (1962). Determination of the base composition of deoxyribonucleic acid from its thermal denaturation temperature. *J Mol Biol* **5**, 109-118.

Masschalck, B.; Deckers, D. and Michiels, C. W. (2003). Sensitization of outer-membrane mutants of *Salmonella typhimurium* and *Pseudomonas aeruginosa* to antimicrobial peptides under high pressure. *J Food Prot* **66**(8): 1360-1367.

Massé, E.; Salvail, H.; Desnoyers, G. and Arguin, M. (2007). Small RNAs controlling iron metabolism. *Curr Opin Microbiol* **10**(2): 140-145.

Mathee, K.; McPherson, C. J. and Ohman, D. E. (1997). Posttranslational control of the *algT* (*algU*)-encoded sigma22 for expression of the alginate regulon in *Pseudomonas aeruginosa* and localization of its antagonist proteins MucA and MucB (AlgN). *J Bacteriol* **179**(11): 3711-3720.

Mattick, J. S. (2002). Type IV pili and twitching motility. *Annu Rev Microbiol* **56**: 289-314.

May, T. B.; Shinabarger, D.; Boyd, A. and Chakrabarty, A. M. (1994). Identification of amino acid residues involved in the activity of phosphomannose isomerase-guanosine 5'-diphospho-D-mannose pyrophosphorylase. A bifunctional enzyme in the alginate biosynthetic pathway of *Pseudomonas aeruginosa*. *J Biol Chem* **269**(7): 4872-7.

REFERENCES

Metzker, M. L. (2005). Emerging technologies in DNA sequencing. *Genome Res* **15**(12): 1767-1776.

Michel, V.; Lehoux, I.; Depret, G.; Anglade, P.; Labadie, J. and Hebraud, M. (1997). The cold shock response of the psychrotrophic bacterium *Pseudomonas fragi* involves four low-molecular-mass nucleic acid-binding proteins. *J Bacteriol* **179**(23): 7331-7342.

Miller, J. H. (1972). *Experiments in Molecular Genetics*. New:York: Cold Spring Harbor Laboratory Press.

Miranda-Ríos, J. (2007). The THI-box riboswitch, or how RNA binds thiamin pyrophosphate. *Structure* **15**(3): 259-265.

Miyoshi, A.; Rosinha, G. M. S.; Camargo, I. L. B. C.; Trant, C. M. C.; Cardoso, F. C.; Azevedo, V. and Oliveira, S. C. (2007). The role of the *vacB* gene in the pathogenesis of *Brucella abortus*. *Microbes Infect* **9**(3): 375-381.

Molina, L.; Ramos, C.; Ronchel, M. C.; Molin, S. and Ramos, J. L. (1998). Construction of an efficient biologically contained *Pseudomonas putida* strain and its survival in outdoor assays. *Appl Environ Microbiol* **64**(6): 2072-2078.

Molina-Henares, M. A.; de la Torre, J.; García-Salamanca, A.; Molina-Henares, A. J.; Herrera, M. C.; Ramos, J. L. and Duque, E. (2010). Identification of conditionally essential genes for growth of *Pseudomonas putida* KT2440 on minimal medium through the screening of a genome-wide mutant library. *Environ Microbiol* (in print).

Moore, P. B. (1995). Molecular mimicry in protein synthesis?. *Science* **270**(5241): 1453-1454.

Mortazavi, A.; Williams, B. A.; McCue, K.; Schaeffer, L. and Wold, B. (2008). Mapping and quantifying mammalian transcriptomes by RNA-Seq. *Nat Methods* **5**(7): 621-628.

Mullis, K.; Faloona, F.; Scharf, S.; Saiki, R.; Horn, G. and Erlich, H. (1986). Specific enzymatic amplification of DNA in vitro: the polymerase chain reaction. *Cold Spring Harb Symp Quant Biol* **51**(Pt 1): 263-273.

Murray, T. S. and Kazmierczak, B. I. (2006). FlhF Is Required for Swimming and Swarming in *Pseudomonas aeruginosa*. *J Bac* **188**(19): 6995–7004.

Nair, B. (2001). Final report on the safety assessment of Benzyl Alcohol, Benzoic Acid, and Sodium Benzoate. *Int J Toxicol* **20**(Suppl 3): 23-50.

Nakashima, K.; Kanamaru, K.; Mizuno, T. and Horikoshi, K. (1996). A novel member of the *cspA* family of genes that is induced by cold shock in *Escherichia coli*. *J Bacteriol* **178**(10): 2994-2997.

Nakazawa, T. and Yokota, T. (1973). Benzoate metabolism in *Pseudomonas putida*(arvilla) mt-2: demonstration of two benzoate pathways. *J Bacteriol* **115**(1): 262-267.

Naughton, L. M.; Blumerman, S. L.; Carlberg, M. and Boyd, E. F. (2009). Osmoadaptation among *Vibrio* species and unique genomic features and physiological responses of *Vibrio parahaemolyticus*. *Appl Environ Microbiol* **75**(9): 2802-2810.

REFERENCES

Nelson, K. E.; Weinel, C.; Paulsen, I. T.; Dodson, R. J.; Hilbert, H.; dos Santos, V. A. P. M.; Fouts, D. E.; Gill, S. R.; Pop, M.; Holmes, M.; Brinkac, L.; Beanan, M.; DeBoy, R. T.; Daugherty, S.; Kolonay, J.; Madupu, R.; Nelson, W.; White, O.; Peterson, J.; Khouri, H.; Hance, I.; Lee, P. C.; Holtzapple, E.; Scanlan, D.; Tran, K.; Moazzez, A.; Utterback, T.; Rizzo, M.; Lee, K.; Kosack, D.; Moestl, D.; Wedler, H.; Lauber, J.; Stjepandic, D.; Hoheisel, J.; Straetz, M.; Heim, S.; Kiewitz, C.; Eisen, J. A.; Timmis, K. N.; Düsterhöft, A.; Tümmler, B. and Fraser, C. M. (2002). Complete genome sequence and comparative analysis of the metabolically versatile *Pseudomonas putida* KT2440. *Environ Microbiol* **4**(12): 799-808.

Neuwald, A. F.; Aravind, L.; Spouge, J. L. and Koonin, E. V. (1999). AAA+: A class of chaperone-like ATPases associated with the assembly, operation, and disassembly of protein complexes. *Genome Res* **9**(1): 27-43.

Nijkamp, K.; van Luijk, N.; de Bont, J. A. M. and Wery, J. (2005). The solvent-tolerant *Pseudomonas putida* S12 as host for the production of cinnamic acid from glucose. *Appl Microbiol Biotechnol* **69**(2): 170-177.

Nishijyo, T.; Haas, D. and Itoh, Y. (2001). The CbrA-CbrB two-component regulatory system controls the utilization of multiple carbon and nitrogen sources in *Pseudomonas aeruginosa*. *Mol Microbiol* **40**(4): 917-931.

Nishikawa, Y.; Yasumi, Y.; Noguchi, S.; Sakamoto, H. and Nikawa, J. (2008). Functional analyses of *Pseudomonas putida* benzoate transporters expressed in the yeast *Saccharomyces cerevisiae*. *Biosci Biotechnol Biochem* **72**(8): 2034-2038.

Nogales, J.; Palsson, B. Ø. and Thiele, I. (2008). A genome-scale metabolic reconstruction of *Pseudomonas putida* KT2440: iJN746 as a cell factory. *BMC Syst Biol* **2**, 79.

Nüsslein, K.; Maris, D.; Timmis, K. and Dwyer, D. F. (1992). Expression and transfer of engineered catabolic pathways harbored by *Pseudomonas* spp. introduced into activated sludge microcosms. *Appl Environ Microbiol* **58**(10): 3380-3386.

O'Hara, E. B.; Chekanova, J. A.; Ingle, C. A.; Kushner, Z. R.; Peters, E. and Kushner, S. R. (1995). Polyadenylylation helps regulate mRNA decay in *Escherichia coli*. *Proc Natl Acad Sci U S A* **92**(6): 1807-1811.

Okamoto, K.; Izawa, M. and Yanase, H. (2003). Isolation and application of a styrene-degrading strain of *Pseudomonas putida* to biofiltration. *J Biosci Bioeng* **95**(6): 633-636.

Oliver, H. F.; Orsi, R. H.; Ponnala, L.; Keich, U.; Wang, W.; Sun, Q.; Cartinhour, S. W.; Filiatrault, M. J.; Wiedmann, M. and Boor, K. J. (2009). Deep RNA sequencing of L. monocytogenes reveals overlapping and extensive stationary phase and sigma B-dependent transcriptomes, including multiple highly transcribed noncoding RNAs. *BMC Genomics* **10**: 641.

Olivera, E. R.; Miñambres, B.; García, B.; Muñiz, C.; Moreno, M. A.; Ferrández, A.; Díaz, E.; García, J. L. and Luengo, J. M. (1998). Molecular characterization of the phenylacetic acid catabolic pathway in *Pseudomonas putida* U: the phenylacetyl-CoA catabolon. *Proc Natl Acad Sci U S A* **95**(11): 6419-6424.

Ottemann, K. M. and Lowenthal, A. C. (2002). *Helicobacter pylori* uses motility for initial colonization and to attain robust infection. *Infect Immun* 70(4): 1984-1990.

Overhage, J.; Lewenza, S.; Marr, A. K. and Hancock, R. E. W. (2007). Identification of genes involved in swarming motility using a *Pseudomonas aeruginosa* PAO1 mini-Tn5-lux mutant library. *J Bacteriol* 189(5): 2164-2169.

Owens, R. M.; Pritchard, G.; Skipp, P.; Hodey, M.; Connell, S. R.; Nierhaus, K. H. and OConnor, C. D. (2004). A dedicated translation factor controls the synthesis of the global regulator Fis. *EMBO J* 23(16): 3375-3385.

Panicker, G.; Mojib, N.; Nakatsuji, T.; Aislabie, J. and Bej, A. K. (2010). Occurrence and distribution of *capB* in Antarctic microorganisms and study of its structure and regulation in the Antarctic biodegradative *Pseudomonas* sp. 30/3. *Extremophiles* 14(2): 171-183.

Parales, R. E.; Ditty, J. L. and Harwood, C. S. (2000). Toluene-degrading bacteria are chemotactic towards the environmental pollutants benzene, toluene, and trichloroethylene. *Appl Environ Microbiol* 66(9): 4098-4104.

Parales, R. E.; Bruce, N. C.; Schmid, A. and Wackett, L. P. (2002). Biodegradation, biotransformation, and biocatalysis (b3). *Appl Environ Microbiol* 68(10): 4699-4709.

Parales, R. E. and Haddock, J. D. (2004). Biocatalytic degradation of pollutants. *Curr Opin Biotechnol* 15(4): 374-379.

Park, S.; Smith, L. T. and Smith, G. M. (1995). Role of Glycine Betaine and Related Osmolytes in Osmotic Stress Adaptation in *Yersinia enterocolitica* ATCC 9610.., *Appl Environ Microbiol* 61(12): 4378-4381.

Pedotti, P.; 't Hoen, P. A. C.; Vreugdenhil, E.; Schenk, G. J.; Vossen, R. H.; Ariyurek, Y.; de Hollander, M.; Kuiper, R.; van Ommen, G. J. B.; den Dunnen, J. T.; Boer, J. M. and de Menezes, R. X. (2008). Can subtle changes in gene expression be consistently detected with different microarray platforms?. *BMC Genomics* 9: 124.

Peix, A.; Ramírez-Bahena, M.-H. and Velázquez, E. (2009). Historical evolution and current status of the taxonomy of genus *Pseudomonas*.., *Infect Genet Evol* 9(6): 1132-1147.

Pfennig, P. L. and Flower, A. M. (2001). BipA is required for growth of *Escherichia coi* K12 at low temperature. *Mol Genet Genomics* 266(2): 313-317.

Phadtare, S. (2004). Recent developments in bacterial cold-shock response. *Curr Issues Mol Biol* 6(2): 125-136.

Phadtare, S. and Inouye, M. (2004). Genome-wide transcriptional analysis of the cold shock response in wild-type and cold-sensitive, quadruple-*csp*-deletion strains of *Escherichia coli*. *J Bacteriol* 186(20): 7007-7014.

Phoenix, P.; Keane, A.; Patel, A.; Bergeron, H.; Ghoshal, S. and Lau, P. C. K. (2003). Characterization of a new solvent-responsive gene locus in *Pseudomonas putida* F1 and its functionalization as a versatile biosensor. *Environ Microbiol* 5(12): 1309-1327.

REFERENCES

Piette, F.; DAmico, S.; Struvay, C.; Mazzucchelli, G.; Renaut, J.; Tutino, M. L.; Danchin, A.; Leprince, P. and Feller, G. (2010). Proteomics of life at low temperatures: trigger factor is the primary chaperone in the Antarctic bacterium *Pseudoalteromonas haloplanktis* TAC125. *Mol Microbiol.* (in print).

Pohl, M.; Sprenger, G. A. and Muller, M. (2004). A new perspective on thiamine catalysis. *Curr Opin Biotechnol* **15**: 335–342.

Poole, K. (2001). Multidrug efflux pumps and antimicrobial resistance in *Pseudomonas aeruginosa* and related organisms. *J Mol Microbiol Biotechnol* **3**(2): 255-264.

Quackenbush, J. (2002). Microarray data normalization and transformation. *Nat Genet* **32**(Suppl): 496-501.

Ramakrishnan, V.; Capel, M.; Kjeldgaard, M.; Engelman, D. M. and Moore, P. B. (1984). Positions of proteins S14, S18 and S20 in the 30 S ribosomal subunit of *Escherichia coli*. *J Mol Biol* **174**(2): 265-284.

Ramos, J. L.; Duque, E.; Huertas, M. J. and Haïdour, A. (1995). Isolation and expansion of the catabolic potential of a *Pseudomonas putida* strain able to grow in the presence of high concentrations of aromatic hydrocarbons. *J Bacteriol* **177**(14): 3911-3916.

Reva, O. N.; Weinel, C.; Weinel, M.; Böhm, K.; Stjepandic, D.; Hoheisel, J. D. and Tümmler, B. (2006). Functional genomics of stress response in *Pseudomonas putida* KT2440. *J Bacteriol* **188**(11): 4079-4092.

Rietsch, A.; Wolfgang, M. C. and Mekalanos, J. J. (2004). Effect of metabolic imbalance on expression of type III secretion genes in *Pseudomonas aeruginosa*. *Infect Immun* **72**(3): 1383-1390.

Ruiz-Lozano and Bonfante (2000). A *Burkholderia* Strain Living Inside the Arbuscular Mycorrhizal Fungus *Gigaspora margarita* Possesses the *vacB* Gene, Which Is Involved in Host Cell Colonization by Bacteria. *Microb Ecol* **39**(2): 137-144.

Régnier, P. and Arraiano, C. M. (2000). Degradation of mRNA in bacteria: emergence of ubiquitous features. *Bioessays* **22**(3): 235-244.

Rühl, J.; Schmid, A. and Blank L. M. (2009). Selected *Pseudomonas putida* strains able to grow in the presence of high butanol concentrations. *Appl Environ Microbiol* **75**(13):4653-6.

Saiki, R. K.; Gelfand, D. H.; Stoffel, S.; Scharf, S. J.; Higuchi, R.; Horn, G. T.; Mullis, K. B. and Erlich, H. A. (1988). Primer-directed enzymatic amplification of DNA with a thermostable DNA polymerase. *Science* **239**(4839): 487-491.

Salzberg, S. L.; Delcher, A. L.; Kasif, S. and White, O. (1998). Microbial gene identification using interpolated Markov models. *Nucleic Acids Res* **26**(2): 544-548.

Sambrook, J.; Fritsch, E. F. and Maniatis, T. (1989). *Molecular cloning: a laboratory manual.* New York: Cold Spring Harbor Laboratory Press.

Sarkar, N. (1996). Polyadenylation of mRNA in bacteria. *Microbiology* **142** (Pt 11): 3125- 3133.

Sarkar, N. (1997). Polyadenylation of mRNA in prokaryotes. *Annu Rev Biochem* **66**: 173-197.

Schurek, K. N.; Sampaio, J. L. M.; Kiffer, C. R. V.; Sinto, S.; Mendes, C. M. F. and Hancock, R. E. W. (2009). Involvement of *pmrAB* and *phoPQ* in polymyxin B adaptation and inducible resistance in non-cystic fibrosis clinical isolates of *Pseudomonas aeruginosa*. *Antimicrob Agents Chemother* **53**(10): 4345-4351.

Segura, A.; Godoy, P.; van Dillewijn, P.; Hurtado, A.; Arroyo, N.; Santacruz, S. and Ramos, J.-L. (2005). Proteomic analysis reveals the participation of energy- and stress-related proteins in the response of *Pseudomonas putida* DOT-T1E to toluene. *J Bacteriol* **187**(17): 5937-5945.

Sharma, C. M.; Hoffmann, S.; Darfeuille, F.; Reignier, J.; Findeiss, S.; Sittka, A.; Chabas, S.; Reiche, K.; Hackermüller, J.; Reinhardt, R.; Stadler, P. F. and Vogel, J. (2010). The primary transcriptome of the major human pathogen *Helicobacter pylori*. *Nature* **464**(7286): 250-255.

Sikkema, J.; de Bont, J. A. and Poolman, B. (1995). Mechanisms of membrane toxicity of hydrocarbons. *Microbiol Rev* **59**(2): 201-222.

Simon, R.; Lam, A.; Li, M.-C.; Ngan, M.; Menenzes, S. and Zhao, Y. (2007). Analysis of Gene Expression Data Using BRB-Array Tools. *Cancer Inform* **3**: 11-17.

Sittka, A.; Sharma, C. M.; Rolle, K. and Vogel, J. (2009). Deep sequencing of *Salmonella* RNA associated with heterologous Hfq proteins in vivo reveals small RNAs as a major target class and identifies RNA processing phenotypes. *RNA Biol* **6**(3): 266-275.

Sonnleitner, E.; Sorger-Domenigg, T.; Madej, M. J.; Findeiss, S.; Hackermüller, J.; Hüttenhofer, A.; Stadler, P. F.; Bläsi, U. and Moll, I. (2008). Detection of small RNAs in *Pseudomonas aeruginosa* by RNomics and structure-based bioinformatic tools. *Microbiology* **154**(Pt 10): 3175-3187.

Sonnleitner, E.; Abdou, L. and Haas, D. (2009). Small RNA as global regulator of carbon catabolite repression in *Pseudomonas aeruginosa*. *Proc Natl Acad Sci U S A* **106**(51): 21866-21871.

Sorek, R. and Cossart, P. (2010). Prokaryotic transcriptomics: a new view on regulation, physiology and pathogenicity. *Nat Rev Genet* **11**(1): 9-16.

Southern, E. M. (1975). Detection of specific sequences among DNA fragments separated by gel electrophoresis. *J Mol Biol* **98**(3): 503-517.

Springer, M. S.; Goy, M. F. and Adler, J. (1979). Protein methylation in behavioural control mechanisms and in signal transduction. *Nature* **280**(5720): 279-284.

Sundin, G. W.; Shankar, S.; Chugani, S. A.; Chopade, B. A.; Kavanaugh-Black, A. and Chakrabarty, A. M. (1996). Nucleoside diphosphate kinase from *Pseudomonas aeruginosa*: characterization of the gene and its role in cellular growth and exopolysaccharide alginate synthesis. *Mol Microbiol* **20**(5): 965-979.

REFERENCES

Terán, W.; Felipe, A.; Segura, A.; Rojas, A.; Ramos, J.-L. and Gallegos, M.-T. (2003). Antibiotic-dependent induction of *Pseudomonas putida* DOT-T1E TtgABC efflux pump is mediated by the drug binding repressor TtgR. *Antimicrob Agents Chemother* **47**(10): 3067-3072.

Tettelin, H.; Masignani, V.; Cieslewicz, M. J.; Donati, C.; Medini, D.; Ward, N. L.; Angiuoli, S. V.; Crabtree, J.; Jones, A. L.; Durkin, A. S.; Deboy, R. T.; Davidsen, T. M.; Mora, M.; Scarselli, M.; y Ros, I. M.; Peterson, J. D.; Hauser, C. R.; Sundaram, J. P.; Nelson, W. C.; Madupu, R.; Brinkac, L. M.; Dodson, R. J.; Rosovitz, M. J.; Sullivan, S. A.; Daugherty, S. C.; Haft, D. H.; Selengut, J.; Gwinn, M. L.; Zhou, L.; Zafar, N.; Khouri, H.; Radune, D.; Dimitrov, G.; Watkins, K.; OConnor, K. J. B.; Smith, S.; Utterback, T. R.; White, O.; Rubens, C. E.; Grandi, G.; Madoff, L. C.; Kasper, D. L.; Telford, J. L.; Wessels, M. R.; Rappuoli, R. and Fraser, C. M. (2005). Genome analysis of multiple pathogenic isolates of *Streptococcus agalactiae*: implications for the microbial "pangenome". *Proc Natl Acad Sci U S A* **102**(39): 13950-13955.

Thiede, B.; Höhenwarter, W.; Krah, A.; Mattow, J.; Schmid, M.; Schmidt, F. and Jungblut, P. R. (2005). Peptide mass fingerprinting. *Methods* **35**(3): 237-247.

Tobe, T.; Sasakawa, C.; Okada, N.; Honma, Y. and Yoshikawa, M. (1992). vacB, a novel chromosomal gene required for expression of virulence genes on the large plasmid of Shigella flexneri. *J Bacteriol* **174**(20): 6359-6367.

Toledo-Arana, A.; Repoila, F. and Cossart, P. (2007). Small noncoding RNAs controlling pathogenesis., *Curr Opin Microbiol* **10**(2): 182-188.

Touati, D.; Jacques, M.; Tardat, B.; Bouchard, L. and Despied, S. (1995). Lethal oxidative damage and mutagenesis are generated by iron in delta fur mutants of *Escherichia coli*: protective role of superoxide dismutase. *J Bacteriol* **177**(9): 2305-2314.

Tsao, M.-Y.; Lin, T.-L.; Hsieh, P.-F. and Wang, J.-T. (2009). The 3-to-5 exoribonuclease (encoded by HP1248) of *Helicobacter pylori* regulates motility and apoptosis-inducing genes. *J Bacteriol* **191**(8): 2691-2702.

Tucker, T.; Marra, M. and Friedman, J. M. (2009). Massively parallel sequencing: the next big thing in genetic medicine. *Am J Hum Genet* **85**(2): 142-154.

Valentin-Hansen, P.; Johansen, J. and Rasmussen, A. A. (2007). Small RNAs controlling outer membrane porins. *Curr Opin Microbiol* **10**(2): 152-155.

Vanderpool, C. K. (2007). Physiological consequences of small RNA-mediated regulation of glucose-phosphate stress. *Curr Opin Microbiol* **10**(2): 146-151.

van der Werf, M. J.; Overkamp, K. M.; Muilwijk, B.; Koek, M. M.; van der Werff-van der Vat, B. J.; Jellema, R. H.; Coulier, L. and Hankemeier, T. (2008). Comprehensive analysis of the metabolome of *Pseudomonas putida* S12 grown on different carbon sources. *Mol Biosyst* **4**(4): 315-27.

Vargas, C.; Argandoña, M.; Reina-Bueno, M.; Rodríguez-Moya, J.; Fernández-Aunión, C. and Nieto, J. J. (2008). Unravelling the adaptation responses to osmotic and temperature stress in *Chromohalobacter salexigens*, a bacterium with broad salinity tolerance. *Saline Systems* **4**: 14.

Veinger, L.; Diamant, S.; Buchner, J. and Goloubinoff, P. (1998). The small heat-shock protein IbpB from *Escherichia coli* stabilizes stress-denatured proteins for subsequent refolding by a multichaperone network. *J Biol Chem* **273**(18): 11032-11037.

Verhoef, S.; Wierckx, N.; Westerhof, R. G.; de Winde, J. H. and Ruijssenaars, H. J. (2009). Bioproduction of p-hydroxystyrene from glucose by the solvent-tolerant bacterium *Pseudomonas putida* S12 in a two-phase water-decanol fermentation. *Appl Environ Microbiol* **75**(4): 931-6.

Wackett, L. P. (2003). *Pseudomonas putida*-a versatile biocatalyst. *Nat Biotechnol* **21**(2): 136-138.

Wai, S. N.; Nakayama, K.; Umene, K.; Moriya, T. and Amako, K. (1996). Construction of a ferritin-deficient mutant of *Campylobacter jejuni*: contribution of ferritin to iron storage and protection against oxidative stress. *Mol Microbiol* **20**(6): 1127-1134.

Wang, N.; Yamanaka, K. and Inouye, M. (1999). CspI, the ninth member of the CspA family of *Escherichia coli*, is induced upon cold shock. *J Bacteriol* **181**(5):1603-9.

Wang, Z.; Gerstein, M. and Snyder, M. (2009). RNA-Seq: a revolutionary tool for transcriptomics. *Nat Rev Genet* **10**(1): 57-63.

Waters, L. S. and Storz, G. (2009). Regulatory RNAs in bacteria. *Cell* **136**(4): 615-628.

Weber, M. H. W. and Marahiel, M. A. (2002). Coping with the cold: the cold shock response in the Gram-positive soil bacterium *Bacillus subtilis*. *Philos Trans R Soc Lond B Biol Sci* **357**(1423): 895-907.

Weinel, C. (2003). *Comparative and functional genome analysis of Pseudomonas putida KT2440*. PhD thesis. Fachbereich Chemie, Universität Hannover, Germany.

Wenzel, S. C.; Gross, F.; Zhang, Y.; Fu, J.; Stewart, A. F. and Müller, R. (2005). Heterologous expression of a myxobacterial natural products assembly line in *pseudomonads* via red/ET recombineering. *Chem Biol* **12**(3): 349-356.

Wery, J.; da Silva, D. I. M. and de Bont, J. A. (2000). A genetically modified solvent-tolerant bacterium for optimized production of a toxic fine chemical. *Appl Microbiol Biotechnol* **54**(2): 180-185.

Whitchurch, C. B.; Beatson, S. A.; Comolli, J. C.; Jakobsen, T.; Sargent, J. L.; Bertrand, J. J.; West, J.; Klausen, M.; Waite, L. L.; Kang, P. J.; Tolker-Nielsen, T.; Mattick, J. S. and Engel, J. N. (2005). *Pseudomonas aeruginosa* fimL regulates multiple virulence functions by intersecting with Vfr-modulated pathways. *Mol Microbiol* **55**(5):1357-78.

Wierckx, N. J. P.; Ballerstedt, H.; de Bont, J. A. M. and Wery, J. (2005). Engineering of solvent-tolerant *Pseudomonas putida* S12 for bioproduction of phenol from glucose. *Appl Environ Microbiol* **71**(12): 8221-8227.

Willenbrock, H.; Salomon, J.; Søkilde, R.; Barken, K. B.; Hansen, T. N.; Nielsen, F. C.; Møller, S. and Litman, T. (2009). Quantitative miRNA expression analysis: comparing microarrays with next-generation sequencing. *RNA* **15**(11): 2028-2034.

REFERENCES

Williams, P. A. and Murray, K. (1974). Metabolism of benzoate and the methylbenzoates by *Pseudomonas putida* (arvilla) mt-2: evidence for the existence of a TOL plasmid. *J Bacteriol* **120**(1): 416-423.

Wood, L. F.; Leech, A. J. and Ohman, D. E. (2006). Cell wall-inhibitory antibiotics activate the alginate biosynthesis operon in *Pseudomonas aeruginosa*: Roles of sigma (AlgT) and the AlgW and Prc proteases. *Mol Microbiol* **62**(2): 412-426.

Wood, L. F. and Ohman, D. E. (2009). Use of cell wall stress to characterize sigma 22 (AlgT/U) activation by regulated proteolysis and its regulon in *Pseudomonas aeruginosa*. *Mol Microbiol* **72**(1): 183-201.

Worsey, M. J. and Williams, P. A. (1975). Metabolism of toluene and xylenes by *Pseudomonas putida* (arvilla) mt-2: evidence for a new function of the TOL plasmid. *J Bacteriol* **124**(1): 7-13.

Wu, C. H.; Apweiler, R.; Bairoch, A.; Natale, D. A.; Barker, W. C.; Boeckmann, B.; Ferro, S.; Gasteiger, E.; Huang, H.; Lopez, R.; Magrane, M.; Martin, M. J.; Mazumder, R.; O'Donovan, C.; Redaschi, N.; Suzek, B.; Xiong, A.; Singh, V. K.; Cabrera, G. and Jayaswal, R. K. (2006). The Universal Protein Resource (UniProt): an expanding universe of protein information. *Nucleic Acids Res* **34**: D187-91.

Yamanaka, K.; Fang, L. and Inouye, M. (1998). The CspA family in *Escherichia coli*: multiple gene duplication for stress adaptation. *Mol Microbiol* **27**(2): 247-255.

Yuki, A. and Brimacombe, R. (1975). Nucleotide sequences of *Escherichia coli* 16-S RNA associated with ribosomal proteins S7, S9, S10, S14 and S19. *Eur J Biochem* **56**(1): 23-34.

Zhang, X.-X. and Rainey, P. B. (2008). Dual involvement of CbrAB and NtrBC in the regulation of histidine utilization in *Pseudomonas fluorescens* SBW25. *Genetics* **178**(1): 185-195.

Zhang, X.-X.; Liu, Y.-H. and Rainey, P. (2010). CbrAB-dependent regulation of *pcnB*, a poly(A) polymerase gene involved in polyadenylation of RNA in *Pseudomonas fluorescens*. *Environ Microbiol* (in press).

Zhu, D.; Cui, S. and Nagata, S. (2008). Isolation and characterization of salt-sensitive mutants of the moderately halophilic bacterium *Salinivibrio costicola* subsp. *yaniae*. *Biosci Biotechnol Biochem* **72**(8): 1977-1982.

I want morebooks!

Buy your books fast and straightforward online - at one of world's fastest growing online book stores! Environmentally sound due to Print-on-Demand technologies.

Buy your books online at
www.morebooks.shop

Kaufen Sie Ihre Bücher schnell und unkompliziert online – auf einer der am schnellsten wachsenden Buchhandelsplattformen weltweit! Dank Print-On-Demand umwelt- und ressourcenschonend produziert.

Bücher schneller online kaufen
www.morebooks.shop

KS OmniScriptum Publishing
Brivibas gatve 197
LV-1039 Riga, Latvia
Telefax +371 686 204 55

info@omniscriptum.com
www.omniscriptum.com

Printed by Books on Demand GmbH, Norderstedt / Germany